A NEW WORLD ORDER

This book begins by identifying a global problematique, a coincidence of four sustained factors: war, insecurity and militarisation; the persistence of poverty; the denial of human rights; environmental destruction. The conventional policy approaches to these problems are analysed through a rigorous critique of the three main United Nations reports of the 1980s, those of the Brandt, Palme and Brundtland Commissions.

Attention is turned from the flawed and partial solutions of these reports to the individuals and organisations involved in policy and action at the grassroots level. Peace and security, human rights, economic development, the environment and human development are all discussed.

The author argues that if the root causes of crisis lie in Western scientism, developmentalism and the construct of the nation state, it is only through the success of democratic popular mobilisation that a new world order, based on peace, human dignity and ecological sustainability, can be created.

The coherence of the theoretical approach and the range of case-studies and profiles – from the Grameen Bank to the Chipko Movement and Petra Kelly to Amnesty International – should ensure a diverse readership from academics and students of development, the environment, war/peace studies and international relations to those involved in new social movements for change.

Paul Ekins has written widely on economics, the environment and development. He is editor of *The Living Economy: A New Economics in the Making*. He is a Research Fellow at the Department of Economics, Birkbeck College, University of London.

A NEW WORLD ORDER

Grassroots movements for global change

Paul Ekins

London and New York

First published 1992
by Routledge
11 New Fetter Lane, London EC4P 4EE

Simultaneously published in the USA and Canada
by Routledge
29 West 35th Street, New York, NY 10001

Reprinted 1993

Printed and bound in Great Britain
by Biddles Ltd, Guildford and King's Lynn.

British Library Cataloguing in Publication Data
Ekins, Paul
A new world order : grassroots movements for
global change.
I. Title
551.6
ISBN 0–415–07114–3
ISBN 0–415–07115–1 pbk

*Library of Congress Cataloging in Publication Data
has been applied for*

All royalties from the sale of this book are payable to the Right Livelihood
Award.

CONTENTS

FOREWORD

by Jakob von Uexkull
Founder and Chairman of the Right Livelihood
Award

Today it is easy to be a pessimist. Despite conferences and action plans, many indicators of human welfare are on a downward trend. The global environment continues to deteriorate. The per-capita income of the poorest countries has fallen in recent years. Many 'expert' predictions thirty years ago – promising us by now a disease-free world of plenty – can be held up to ridicule.

For those so inclined it is also still possible to be an optimist, believing that the next scientific/technological fix will usher in paradise on earth. Nuclear fission might have developed a few unforeseen problems but nuclear fusion will finally provide energy too cheap to meter. Biotechnology will correct nature's insufficiencies and genetic engineering will produce a race of healthy, optimistic humans. Past Cassandras can be rubbished for their mistaken predictions and their concerns about limits to growth attributed to 'misguided, middle-class anxiety' (*The Sunday Times*, 13 January 1991).

So either nothing can be done – or everything will be all right if we stay on course. But there is a third less complacent but more exciting alternative: like William James, we can be possibilists. To be a possibilist means accepting the very serious diagnosis of the state of our world now agreed on, not just by 'radicals' but right up to the top of our international organisations. Mostafa Tolba, Director-General of the United Nations Environmental Programme, simply says that the life-support systems of our planet are collapsing. It means challenging at every turn the Julian Simons of this world who are still 'disputing the various alarming estimates of tropical deforestation, species extinction, eroding topsoil, paved over farmland and declining fisheries' (*The Guardian*, 28 December 1990). But being a possibilist also means rejecting the self-fulfilling pessimism of those who tell us that we must of necessity pollute our environment, poison ourselves and consume the future for the sake of short-term greed and comfort, because such is human nature.

This book provides ammunition for possibilists. If the pessimists were right, the work described would be 'contrary to human nature'. If the optimists were right, it would be irrelevant. But these initiatives are not picturesque niches to escape to in case things go wrong. They are part of another global trend, clearly incompatible with the conventional wisdom of both optimists and pessimists. They force us to make a choice.

For you cannot *both* poison and save the environment, you cannot simultaneously empower the people *and* the multi-national companies, you cannot combine local self-reliance with the primacy of global 'free' trade. The major components of the present order identified by Paul Ekins – scientism, industrial developmentalism and nation-statism – tolerate no other gods. It is easy to believe that we can have the best of all worlds because the dominant system is flexible enough to pay tribute to almost all the 'alternative' paths described here. But the gap between theory and practice grows wider daily. The Brundtland Report has been hailed, despite its contradictions, as the most radical document to come out of a grouping consisting of representatives of the global elites. But Mrs Brundtland herself is now working hard to make Norway a member of the EC, an entity which stands for much of the short-sighted growthism condemned in her report. When challenged on this point by a Danish MEP, 'all she replied was that the single market is really a unique opportunity for Norwegian trade and business' (*Notat*, Copenhagen, 25 November 1988). The World Bank is another example, as this book describes, of more and more bank-financed ecological horror stories combined with the proliferation of new environmental staff, policies and action plans.

The dominant path occupies so much space and absorbs so many resources that alternatives rarely stand a chance. Funding to develop renewable energy resources continues to be heavily squeezed by the old commitments to nuclear power. In the Third World 'bad aid' destroys self-sufficiency and thus the basis for 'good aid'. Quite often the major costs of changing course are not the environmentally sound, less disruptive and much cheaper alternatives, but the costs of 'counter-development', i.e. the work needed to block – and sometimes dismantle – the huge, harmful products of bad aid.

In many countries, to be seen as an 'enemy of progress' is not just a recipe for ridicule but a threat to life. Cultural imperialism has been so successful that valuable indigenous initiatives in the South are often completely blocked unless they have been recognised in the North. That is why the Right Livelihood Award ('the Alternative Nobel Prize') was established – to provide an alternative to all those honours which only recognise what falls within the paradigm of the modern Western worldview. Also, if our most important needs today are not new technological fixes but new social values and institutions, making better use of knowledge we already have, then these priorities need to be reflected in what our society honours and supports. Professor Hans-Peter Dürr, one of the Right Livelihood Award recipients profiled in this book, has often remarked on the very different possibilities of obtaining funding for his various activities. While his work on elementary particle physics has never been short of money – often from sources which understand very little or nothing of what he is doing – his peace research and environmental and social innovation work have always been starved of funds.

For years it has been clear that most people, even in the rich countries, would prefer a 'greener', gentler world, want greater control over their lives, and fairness for the poor and future generations. But how do we get there? From the 'experts', little can be expected. Futurists Joseph Coates and Jennifer Jarratt in a recent study of the work of major futurists conclude that

> Potential impacts of the greenhouse effect and other global environmental problems are not given detailed attention. The futurists barely ask who will pay the cleanup costs. Little is provided on how to ensure a livable environment for the future. We are told that we need technological answers on how to alleviate environmental problems before they become irreversible, but the structural and economic changes that may be necessary are not explored.
>
> (Coates and Jarratt 1989).

So, as in previous periods of confusion and doubt, when traditional institutions lose credibility and the future is under threat, it is up to those whom Kropotkin called the 'little people' to find new ways to work together to meet the needs of our time and circumstances. The existence of the Right Livelihood Award has helped bring into the open some of these grassroots expressions of humanity's irrepressible, instinctive urge for mutual aid. I am very happy to see so many of their stories published here. I am sure this book will help not only to spread their message but also to create the understanding which will remove the stones placed in their path – for their future is our future.

Jakob von Uexkull

ACKNOWLEDGEMENTS

In thanking those who have contributed to the realisation of this book, it is with great gratitude that I acknowledge my principal helpers and researchers, for they are none other than the people on whose work this book is largely based. Without their work not only would this book not have been written, but there would have been nothing to write about. I am immensely grateful therefore, firstly that they exist at all; secondly that, when asked, they cheerfully took time off from their hectic schedules to assist my labours; and thirdly that they expressed both encouragement and enthusiasm for what I was trying to do.

I must also thank the Right Livelihood Awards Foundation, its Founder and Chairman Jakob von Uexkull and its Trustees, for giving me the opportunity as Research Director of the Foundation to explore the fascinating and inspiring world of the nominations for the Award, which form the core material of this book. The Research Fellowship the Foundation endowed at Bradford University's School of Peace Studies gave me three rewarding years from 1987 to 1990 during which the basic research for this book was done.

Many other people gave me valuable help and advice as to how to organise into a coherent form the great mass of disparate material for this book; without them the book would be much worse than it is. My thanks to them all. Such errors as remain should, of course, be laid at my door alone.

Paul Ekins

INTRODUCTION

In the late twentieth century it is now increasingly recognised worldwide that humanity faces four interlocking crises of unprecedented magnitude, all of which have the potential for the destruction of whole peoples and some of which threaten the extinction of the human race itself. The four crises are:

- the existence and spread of nuclear and other weapons of mass destruction and the overall level of military expenditure;
- the affliction with hunger and absolute poverty of some 20 per cent of the human race, mainly in what is misleadingly called the Third World;
- environmental pollution and ecosystem and species destruction at such a rate and on such a scale that the very biospheric processes of organic regeneration are under threat;
- intensifying human repression resulting from the increasing denial by governments of the most fundamental human rights and the inability of increasing numbers of people to develop even a small part of their human potential.

As one would expect from the enormity and range of the above problems, each one has spawned its own enormous literature of analysis, prescription and case-study. It is neither possible here, nor is it the purpose of this book, to provide a comprehensive insight and review into this literature. The works cited will, however, give those who want such comprehensiveness ample entry points into the literature of their choice.

The purpose of Chapter 1 is firstly, and very briefly, to make the case that each of the issues mentioned does in fact amount to a present catastrophe and future threat of enormous proportions; secondly, to outline equally briefly the root causes of these crises; and thirdly, to indicate the relations between the crises themselves and between their causes. Chapter 2 will explore principal conventional responses to the crises, asking why they do not in the main even seem to be containing, far less reversing, the chief trends. From this analysis will come the perception that new approaches are needed to these problems, and some of the desirable theoretical properties of the approaches will be outlined.

Chapters 3 to 8 will show that in many fields such approaches already exist. In some cases they have been operative for a consider-able time and have achieved significant practical success. Most of the case-study material which will be presented in these chapters comes from the nomination files of the Right Livelihood Award, of which the author has been Research Director since early 1987. This Award, started in 1980 and widely known as the Alternative Nobel Prize, was founded explicitly to address the crises identified earlier, and to honour those who seemed to have found solutions on a personal or organisational level which ameliorated one or more of the crises without intensifying the others. Since its foundation the Award has received over 300 nominations including many of the most innovative and beneficial social and economic projects in the world today. It is from this largely unknown database that the great majority of the quoted examples are drawn. Further details about the Award, its recipients and other publications about it are given in Appendix 1.

It must be emphasised that the case-studies themselves are of initiatives selected from a large potential universe of which the Right Livelihood nominations only comprise a very small fraction. The initiatives cannot be described as the 'best' in this universe nor are they in any real sense 'representative', for a prime characteristic of these sorts of initiatives is their diversity. Rather these initiatives have been chosen because:

- They seem to have defined the problem they are addressing in a coherent and realistic way;
- Their solution to it seems to go to the root of the problem;
- They have met with significant success;
- Their approach enables broader inferences to be drawn, which are of relevance to other problem areas and their solutions.

Thus the case-studies have been used to build up a patchwork of overlapping approaches to the global problematique described in Chapter 1. The pattern thus created, unlike the flawed and partial solutions discussed in Chapter 2, provides a framework within which the problematique can be addressed. This pattern will be revealed in Chapters 3 to 8, showing that there are many common threads and themes to the case-studies, which cover a very wide range of subjects, even when they seem to be addressing quite different problems in different fields. Such a broad framework, and an eclectic interdisciplinarity in describing it, seems necessary in view of the pervasiveness and interdependence of the issues addressed. It is also in line with some recent academic thinking, such as Johan Galtung's (1985) approach to peace research, based on a perception that there are four broad 'spaces' of relationships which produce peace or violence: the human space of personal relationships; the social space, including culture, politics and the economy; the global space of world systems; and the ecological space comprising the relationship of people with nature. In ranging over this whole terrain, this book seeks to show that these spaces are connected in a wide variety of ways and that successful approaches in one space are of great relevance to another, whence they can contribute greatly to the betterment of the human condition as a whole. Pulling this conclusion together is the task of Chapter 9.

1

THE GLOBAL
PROBLEMATIQUE

THE MILITARY MACHINE

The essential statistics of military power are readily accessible.
There are three principal categories of weapons of mass destruction,
nuclear, chemical and biological. The nuclear category has received
the most public attention and its extent can be briefly summarised.
Five nations – China, France, the UK, US, and USSR – admit
to being nuclear-armed, though several other countries, including
Israel and South Africa, are known to have nuclear weapons, and
several others, including Brazil, India, and Pakistan are thought
to have a nuclear capability. Proliferation of nuclear weapons is
likely to spread even more widely if civil nuclear reactors continue
to be constructed in otherwise non-nuclear countries. There is an
abiding danger that the Nuclear Non-Proliferation Treaty, which
up to 1990 had remained practically a dead letter in terms of the
nuclear disarmament it envisaged, will become one with regard to
non-proliferation also.

The vast majority of nuclear weapons are held by the two
superpowers, the US and USSR, who in 1989 were thought to
have some 55,500 warheads between them, 98 per cent of a world
total which has a combined power of 1,200,000 Hiroshima-type
bombs (Sivard 1989, p.9). It is estimated that this amount, 1,000
times the firepower used in *all* wars since the introduction of
gunpowder six centuries ago, could destroy the planet at least
twelve times over (Sivard 1986, p.7). While it might be thought
that this overkill represented the apogee of nuclear deterrence and
that efforts might therefore be turned to maintaining rather than
increasing or improving this destructive power, this is far from being
the case. Arms levels grew inexorably through the 1980s (at least until

the accession of Mr Gorbachev). The US in particular still strives to improve the stealth and accuracy of its nuclear weapons, against all deterrence logic. In proceeding with SDI (the 'Star Wars' Strategic Defense Initiative), which by 1989 had cost US$17 billion (Sivard 1989, p.15), the US continues to signal to the USSR and the world in general that it wishes to obtain invulnerability from nuclear attack and so break the balance of deterrence in its favour. It is not just the USSR that has legitimate cause for concern if the US became able to deploy its firepower free from risk of retaliation.

While nuclear weapons have long attracted great public interest and debate, chemical weapons have recently become an increasingly serious source of concern. The proven use of these weapons by Iraq in the Iran – Iraq war and the general perception of that use as having been both militarily successful and relatively costless politically, have caused chemical weapons to be widely perceived as 'the poor countries' nuclear weapon', due to their relative ease and cheapness of manufacture. Prospects for a worldwide ban on the development, production and possession, as well as use, of chemical weapons, a convention which has been in preparation since 1969 by the forty-nation Conference on Disarmament in Geneva, have accordingly receded. Unless agreement is reached quickly, it is likely that even medium to small states in the Third World will either have acquired or be able quickly to acquire chemical weapons should they wish to do so.

One of the very few significant measures of arms control before the recent US/USSR agreement on intermediate range nuclear weapons was the Biological Weapons Convention of 1972, ratified by 105 nations, which forbade the development, production and stockpiling of these weapons, as well as ordering the destruction of existing stocks. However, under the guise of research into defence against biological weapons, up to ten countries are now thought to have potentially offensive research programmes under way in this field. Advances in biotechnology and genetic engineering have enormously increased the potential for the development of lethal organisms, and it seems only a matter of time before these too become routine, if unadmitted, components of many arsenals.

While the weapons of mass destruction occupy the high-profile end of the military spectrum, they account for only a small proportion, less than 20 per cent (Sivard 1989, p.33) of global military spending, which in 1988 amounted to approximately US$1,000 billion (Sivard 1989, p.12). This sum buys not only military hardware. It employs well over half a million scientists and engineers in military research and development. It puts some 25 million people into uniform. It is hard to envisage the opportunity cost of US$1,000 billion except by reference to the estimated cost of various international programmes which have been devised to tackle major generally-recognised problems in other areas: the FAO/UNDP/World Bank/World Resources Institute Tropical Forestry Action Plan, US$8 billion over five years; the UN Water and Sanitation Decade's plan of action to bring clean water to every citizen in the world by the year 2000, US$300 billion over the 1980s; the 1977 inter-governmental plan to halt the spread of deserts, US$90 billion over the last two decades of the century (MacNeill *et al.* 1989, p.10); UNICEF's estimate of 'the additional cost of meeting the most essential of human needs' of everyone on earth, a maximum of US$500 billion over ten years (UNICEF 1989, p.65); extending the World Health Organization's Expanded Programme on Immunization to all Third World children, US$2 billion (Sivard 1989, p.33). It can be easily seen that these large programmes, none of which has yet been anywhere near implemented, would all fit comfortably within one year's global arms expenditure.

If the military expenditure could be shown to have prevented war, then at least it might be seen to have some justification. Unfortunately this is far from being the case. From 1945–89 some 127 wars occurred round the world, killing an estimated 22 million people and injuring countless others (Sivard 1989, pp.11, 22). In 1988 some 27 wars (annual deaths exceeding 1,000) were being fought (Sivard 1987, p.28). Because most of them were being fought with modern imported weapons they were much more destructive than if each country had had to rely on its own weapons capability. The global trade in arms, principally supplied by France, the UK, US and USSR, with Brazil and India being increasingly prominent Third World arms manufacturers, ensures that the most modern means of killing people are quickly accessible to any government in the world.

It is often argued that nuclear deterrence has, at least, kept the peace in Europe since 1945 and prevented nuclear war. The argument is also heard that the recent superpower détente and agreement was due to NATO's bargaining from strength, by which is meant its decision in 1979 to deploy Cruise and Pershing missiles.

Both these arguments exhibit the *post hoc ergo propter hoc* logical fallacy. Just because one event succeeds another, it does not follow that it was caused by it. It is just as logical to assert that forty years of European peace have been achieved and that nuclear war has so far been avoided *in spite* of the policy of nuclear deterrence, with all its attendant risks of bluff, brinkmanship, misunderstanding and mistake. Similarly it can just as logically be argued that Mr Gorbachev was only able to gain support within the USSR for his arms control proposals because of the huge manifestation of public opinion against the Cruise and Pershing decisions in both Europe and the US, which reassured the Soviets that despite the tough NATO stance, Western publics were deeply desirous of an end to the Cold War, as has subsequently been proved by Mr Gorbachev's popularity with those publics.

The root cause of wars and the arms race is as old as recorded history itself: aggression, on the one hand, and consequent insecurity on the other. Some countries acquire arms to attack others. Others acquire them for defence. Most claim the latter motivation but ensure that they purchase arms which will serve either purpose, either because of secret aggressive intent or because they genuinely believe that the best form of defence is attack.

Security acquired through weapons is, at best, a zero-sum good: one side's greater security is another's increased vulnerability (defence-only weapons can invalidate this equation, but they are still very much the exception rather than the rule). This is what puts the spiral into the arms race as each country seeks its own security at another's expense. After each round, security is likely to have diminished at higher cost, because both sides are more heavily armed with more deadly weapons. An opportunity cost of the weapons is a missed chance to strengthen both country's economic and social infrastructure.

It is extremely unlikely that, in a world where nuclear, chemical and biological weapons are commonplace, they will not be widely and catastrophically used sooner or later. The insecurity engendered by large arms arsenals, intertwined with the other crisis-insecurities of mass poverty, ecological destruction and population repression, and combined with all the other factors of race, ideology, expansion-ism and religious belief which have traditionally led to conflict, all mean that wars in today's world are likely to increase in number and intensity for as long as security is perceived as emanating from the barrel of a gun.

THE HOLOCAUST OF POVERTY

Poverty is both an absolute and relative phenomenon. The absolute variety afflicts about one billion people, the 20 per cent of the world population that is not able on a regular basis to satisfy their most basic human subsistence needs. In consequence the lives of these poor tend to be short. UNICEF has estimated that 15 million children die each year from such poverty.

If absolute poverty is basically a physical condition, relative poverty is more a function of expectations and opportunity in a particular society. A basket of goods or skills which, in one society, may confer self-respect and social worth and usefulness may, in another, bring only alienation, isolation and despair. The psychical component of relative poverty has special implications for attempts to relieve it as will become clear.

Whereas the great majority of the absolutely poor live in the Third World, the relatively poor are concentrated more in industrial countries. In the UK in 1987 over 10 million people, (nearly a fifth of the population, were living in poverty (Oppenheim 1990, p.1). Few of these if any will be actually starving. Most will be relatively poor. There is plenty of evidence that relative poverty is as excruciating in its own way as the absolute variety. Its influence can cause people to turn to drugs, violence and crime. It breaks up families, destroys sociability and induces profound personal stress.

As with war, poverty, especially of the absolute kind, seems long to have been an integral part of the human condition. The difference between poverty and war, and the disappointment, lies in the fact that for at least the past 100 years in industrial countries, and forty years in the so-called 'developing' world, economic growth is supposed to have been leading inexorably to the abolition of poverty. That is what the whole project of 'development' was all about.

The reality has been quite different. There is little evidence that absolute poverty is generally on an established downward trend. In many countries of the Third World, especially in Latin America and Africa, it is clear that the reverse is the case. For example, in Sub–Saharan Africa:

- per capita incomes have fallen by almost 25% during the 1980s;
- investments have fallen by almost 50% and are now in per

capita terms lower that they were in the middle of the 1960s;
- imports are today only 6% of per capita imports in 1970;
- exports have fallen by 45% since 1980;
- the external debt has grown from $10 billion in 1972 to a staggering $130 billion in 1987. The capacity to service it has not kept pace, thus creating an unmanageable situation for some twenty low-income countries;

. . .

- the share of children starting school is beginning to decline and in some places the infant mortality rate is on the increase.
(Swedish Ministry of Foreign Affairs 1988, p.11)

In many cases it is not lack of 'development' that has brought popular impoverishment, but 'development' itself, as when natural resources that provide a decent subsistence livelihood for large numbers of people are turned into industrial raw materials that benefit relatively few. Those who are displaced or dispossessed or both probably began neither absolutely nor relatively poor. While modest, their lives are likely to have been both self-sustaining and commensurate with their expectations and those of their peers. They will have been useful, productive members of their society.

After the 'development' project, be it large dam, plantation, logging, or industrial fishery or whatever, these people will be both absolutely and relatively poor: ill-fed, ill-clothed, and ill-housed and transplanted from their traditional society into one of quite different values and priorities in which it is extremely difficult for them to participate. An excellent current example of this tragic process can be seen in the Narmada Valley Project in India's States of Gujarat, Madhya Pradesh and Maharashtra, described in some detail in Chapter 5. The two largest dams involved in this project will between them displace over 200,000 people, many of them tribals, who will have their villages and fields totally or partially flooded. Despite state government rhetoric, there is not the remotest prospect that sufficient fertile land will be found elsewhere to give these people independent livelihoods. Cut off from their traditions, their communities destroyed, the vast majority of the displaced will drift to the slums of Bombay or Ahmedabad, like countless development refugees before them. Such rural–urban migration has historically been a common factor of all industrialisation. The agricultural beneficiaries from the dam, at least before it silts up and before the salinisation of irrigated land, will be predominantly existing farmers and landowners, while the urban middle and industrial classes will be the main users of the hydro-electricity. There is ample evidence that wealth accruing to these social groups in developing countries does not trickle down to the poor, beyond the relatively small amount of direct employment that is generated. Rather the income differentials between the landless and dispossessed, and the rest, are increased and poverty intensified. As for drinking water, another supposed benefit of the dam, there are grave doubts that the project will even deliver this resource over the hundreds of kilometres required to reach the poorest peoples in the most arid lands, who are often cited as the project's main justification.

One of the reasons why developing country wealth does not reach the poor is that those who control it tend to be more interested in Western markets and Western lifestyles. Their purchase of Western armaments and consumer goods does nothing to give livelihoods to their own people. Paradoxically, therefore, 'development' can end up giving more benefit to the rich countries which provided the initial 'aid' than to the poor people in the countries which received it. This sort of development cycle can be pictured as in Figure 1. It is because of this sort of cycle that increasing numbers of grassroots activists in Southern countries are regarding 'development' as dangerous to and exploitative of poor people in poor countries.

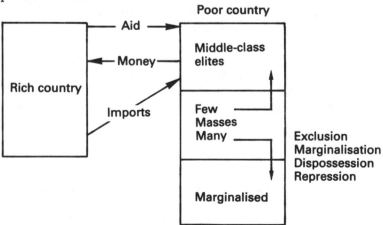

Figure 1: The 'aid and development' cycle

THE ENVIRONMENTAL CRISIS

For most of the past ten years it has been necessary to argue forcefully against general scepticism that there is in fact a serious environmental crisis of global dimensions, but public awareness has changed dramatically since about 1987 so that general concern for the environment has now forced the issue near the top of the international political agenda.

As with armaments and poverty, the problem is quite simply stated: the ability of the biosphere to support and sustain human, animal and plant life is being severely eroded by the numbers of people on the planet and the impact of their economic activities.

Standard environmental economic analysis now recognises that the environment contributes to human life in three fundamental ways. It provides resources for the human economy; it disposes of the wastes from that economy; and it directly provides services for people in many different ways, some of an aesthetic nature by way of amenity (e.g. countryside, scenery, wilderness), and some that are crucial to survival, (e.g. regulation of the climate and atmosphere). It is interesting to note that the first focus of environmental concern was predominantly on the first of these environmental functions, the provision of resources. The Club of Rome's landmark publication *Limits to Growth* (Meadows *et al.* 1972) raised questions about the future supply of minerals and fossil fuels at pertaining (and increasing) levels of consumption. Fifteen years later the attention is much more on the limited ability of the environment to absorb the wastes of the economic process, and the consequent impact of those wastes on the great life-regulating biosystems and on human health. The now familiar litany of pressing problems includes global warming through the emission of the so-called greenhouse gases (carbon dioxide, methane, nitrous oxide and chlorofluoro-carbons [CFCs]), acidification of lakes and forest die-back due to the emission of sulphur dioxide, nitrogen oxides and ozone, depletion of the ozone layer by the emitted CFCs and the generation of increasing quantities of hazardous (chemical or radioactive) wastes.

Interacting with these waste-disposal effects is the ongoing global destruction of bioresources, expressed, though not exhaustively, in such words as deforestation, desertification and species extinction. More than 40 per cent of the world's original 6 million sq. miles of tropical forests have already disappeared and continue to do so at a rate of 30,000–37,000 sq. miles every year, with a similar area suffering from gross degradation (Myers 1986a, p.9); 6 million hectares are added to the world's deserts each year, with an area larger than Africa and inhabited by 1 billion people now at risk; topsoil is being lost at an estimated 25 billion tons per year, roughly the amount covering Australia's wheatlands (MacNeill *et al.* 1989, p.5); and Norman Myers has estimated that, at present rates, hundreds of thousands of species will be extinguished over the next twenty years from only 400,000 sq. miles of prime tropical forest (Myers 1986a, p.12). The ability of the biosphere to sustain such damage into the future without major climatic or other upheavals is to be doubted.

THE DENIAL OF HUMAN RIGHTS

Amnesty International (AI) reported that in 1988 it investigated or adopted people as Prisoners of Conscience (those imprisoned solely because of their belief, sex, ethnic origin, language or religion and who have neither used or advocated violence) in eighty-four countries (Amnesty International 1989, p. 17). In the same year AI brought to the attention of the UN Special Rapporteur on Torture cases from forty-three countries (ibid. p.23). Ruth Leger Sivard's research indicates that there are sixty-four military regimes in the Third World (Sivard 1989, p.21).

There are no simple reasons for human repression on this scale just as there are no simple solutions to it. Certainly the situation is exacerbated by the three problems discussed earlier – war and the arms race, poverty and maldevelopment, and environmental destruction – and by the streams of refugees to which each of these problems give rise. But equally it can be said that there has been a massive failure on behalf of the most democratic nations of the world to give practical effect abroad to their widely professed rhetoric in support of human rights, especially when such effect would have conflicted with their economic or strategic interests. While all Western countries have collaborated with and supported oppressive regimes to some extent, the clearest examples are probably seen in the actions of the US in Central and Latin America and elsewhere. In Chile and Nicaragua, the US made it absolutely clear that it would not tolerate regimes that were unco-operative with US interests; in Brazil and the Philippines it indicated just as clearly that, provided such co-operation was forthcoming, no military junta or dictator could be too brutal to qualify for US support. It is a squalid, hypocritical record that must have given much comfort to repressors the world over.

No single event could have illustrated more clearly the linkage between peace, human rights and environmentally sustainable development than the Gulf War. Saddam Hussein's ruthlessness, bloodiness and violations of international law are not in dispute. Yet he is the same man whom Western powers assiduously supported during the Iran – Iraq War; to whom Western powers sold practically any technology for which he asked to allow him to build the fourth largest army in the world, replete with weapons of mass destruction in a geographical region long recognised to be a potential powder-keg; and whose use of chemical weapons against his own unarmed Kurdish citizens raised barely a whisper of protest, and certainly no sanctions, from the international community. Had the West been half as resolute and principled then on behalf of the Kurds, when Kuwait's oil was not at issue, as it claimed to be during the Gulf War, when oil was widely perceived as the principal reason for its fine words and general bellicosity, Kuwait would probably not have been invaded and hundreds of thousands of Iraqis and Kurds would still be alive. And the US would not have needed either Iraq's or Kuwait's oil if it had taken steps in the 1980s to, for example, raise the average mpg of its car fleet from 18 to 21 (RMI 1990, p.3). New cars can now routinely average more than 30 mpg.

It is not just that the four crises – militarisation, poverty, environmental destruction, human repression – are interlinked, although even taken singly they would be difficult enough to ameloriate, let alone resolve. The situation is made far more intractable by the fact that each of the problems actually reinforces the others. War causes poverty, environmental damage and repression. Poverty causes environmental damage and can lead to revolt, as can repression. Environmental damage causes poverty but it is also true that many 'development' projects which are supposed to relieve poverty also cause environmental damage, as the example of the Narmada dams makes abundantly clear. So does much economic growth in industrial countries, which is also often perceived to be necessary for world development. It is clear that these interactions turn the four separately described problems into a single, systemic global problematique of great complexity.

2

THE NEED FOR NEW
APPROACHES

THREE UN COMMISSIONS

It has taken approximately twenty years for the full enormity of
the global problematique to become apparent. Many of the first
signs of the gathering crisis can, with hindsight, be clearly sighted
in the late 1960s and early 1970s; the loss of US innocence with
the Vietnam War; the breakdown of the Bretton Woods currency
accord in 1973; the Stockholm Conference on the Environment in
1972, with *Limits to Growth* (Meadows *et al.* 1972) the same year;
the Pearson Report of 1969 sounding the first doubts about aid-led
development (Pearson *et al.* 1969).

None of these early alarm bells led to any thorough rethinking
and business as usual prevailed. During the 1970s many countries in
the newly independent Third World became the battleground of the
superpowers' hostile ideologies, and a major repository for their and
their allies' weapons. The Cold War went into a new freeze, and the
arms race a new spiral, following the 1979 Cruise/Pershing decision,
President Reagan's 'evil empire' rhetoric and the Soviet invasion
of Afghanistan. Thinking on economic development sought to
broaden out somewhat from its extremely narrow focus on financial
indicators, culminating in the basic needs approach towards the end
of the decade, but any practical effect of these ideas was drowned
in the sea of Third World debt contracted during this decade,
and the dramatic loss of confidence in multilateral development
co-operation thereafter. As for the environment, the demolition of
Limits to Growth by the economics profession put that issue back
to sleep, until 1980 when the World Conservation Strategy (IUCN
1980) started raising the issues about biodiversity and pollution
damage which are those on the agenda today.

Fittingly, perhaps, it was the United Nations that first seemed to grasp the full global dimension of the developing situation and established over six years three independent commissions to report on different aspects of what was coming to be perceived as a common crisis: the Brandt Commission (Independent Commission on International Development Issues), established in 1977, which published two reports, *North–South, a Programme for Survival* (BR1–ICIDI 1980) and *Common Crisis* (BR2–Brandt Commission 1983); the Palme Commission (Independent Commission on Disarmament and Security Issues) established in 1980, which reported two years later with *Common Security: A Programme for Disarmament* (ICDSI 1982) and the Brundtland Commission (World Commission on Environment and Development) established in 1984, whose *Our Common Future* appeared in 1987 (WCED 1987).

As one would expect from these eminent, well resourced Commissions, comprising in the main committed, sincere and well-informed people, their reports were authoritative statements of the problems they were examining and their conclusions worthy of close study. Without exception they perceived the situation on which they were focusing to be one of crisis or even a question of survival.

Palme Report

Looking into the future, considering the potential long-term effects of many nuclear explosions on the human gene pool and the incidence of cancer, to say nothing of the effects on the ozone layer and resultant destruction of animal and plant life and eventual climatic changes, human life itself could be in jeopardy. Thus humanity would face the ultimate risk – its own extinction.

(p.52)

Brandt Reports

The first Brandt Commission Report, (BR1–ICIDI 1980) was so aware of the momentous nature of the issues under discussion that it was subtitled 'A Programme for Survival', and Willy Brandt, the Commission's Chairman, introduced it with the words: 'Our Report is based on what appears to be the simplest common interest: that mankind wants to survive. . . . If reduced to a simple denominator, this Report deals with peace. . . . There is a growing awareness

15

that an equal danger (to war) might be chaos – as a result of more hunger, economic disaster, environmental catastrophes and terrorism' (BR1 p.13). By the time of the second Brandt Report (BR2–Brandt Commission 1983) three years later the Chairman's warning had grown starker: 'Our situation is unique. Never before was the survival of mankind itself at stake' (BR2 p.9).

Brundtland Report

There are environmental trends that threaten to radically alter the planet, that threaten the lives of many species upon it, including the human species.

(p.2)

Nature is bountiful but it is also fragile and finely balanced. There are thresholds that cannot be crossed without endangering the basic integrity of the system. Today we are close to many of those thresholds; we must be ever mindful of the risk of endangering the survival of life on Earth.

(pp.32–3)

THE PALME REPORT

The Palme Commission report makes its recommendations after an analysis which repeatedly stresses its perception of the realities of defence and security in the nuclear age. Its chief cause for concern is that these realities were becoming clouded by the dynamics of the arms race. The realities include the following:

- There are no effective defences against missiles armed with nuclear warheads; none exist and none are likely to be developed in the foreseeable future (p.5).
- Nuclear war is unwinnable by any participant. Its effects would render all sides effective losers, 'an unprecedented catastrophe for humanity and suicide for those who resorted to it' (p.141).
- The concept of limited nuclear war is fundamentally unsound. Its escalation to all-out nuclear exchanges is practically inevitable (e.g. pp.43ff).
- 'Nuclear deterrence could collapse in many different ways. . . . (It) cannot provide the long-term basis for peace, stability and equity in international society' (p.142). Three factors tend to undermine its efficiency as reliance on it is prolonged:

16

insensitisation of policymakers to the dangers of nuclear weapons; technological advances, making concepts of 'first strike' or 'winnable' or 'limited' nuclear war more plausible; the increased probability of accidents or escalation through crisis (e.g. p.41).
• Expenditure on weapons, especially nuclear weapons, no longer automatically increases security; it now tends to reduce it (p.4).

On the basis of these tenets, the Palme Commission report builds its notion of common security, involving the renunciation of any quest for unilateral military advantage and an acceptance of parity 'to establish security at the lowest possible level of armaments' (p.9). The Commission's recommendations to achieve this common security fall into six sections:

1. Curbing the East–West arms race and improving East–West relations;
2. Preventing a new generation of nuclear weapons qualitatively more sophisticated than at present;
3. Building confidence between potentially hostile states, through verification and other measures;
4. Strengthening the United Nations security system;
5. Developing regional programmes for common security;
6. Recognising the importance of economic security.

The most potent idea in the Palme Commission report is that 'military strength alone cannot provide real security' (p.4). Yet its subsequent analysis is overwhelmingly concerned with military strength, although one of the six chapters is devoted to 'The economic and social consequences of military spending' (pp.71ff). This seems to represent a missed opportunity to develop a distinctive new theory of national and international security which would take full account of the new dimensions of national security and global interdependences which the Palme Report acknowledges but never really integrates into its analysis.

Such a new approach to security would need to include at least the following four dimensions: military security; civil security; economic security; ecological security. These 'securities' are obviously interlinked. In particular, ecological security promotes economic security which promotes civil security. Militarisation of the economy, however, in the quest for military security undermines domestic economic security (Brown *et al.* 1986; Humm 1985; and Melman 1986, all show that strong domestic military production is

17

actually an economic liability). Spending to enhance security should, therefore, be spread over each of these areas in order to maximise the security return.

Military security

This does not only reside in being able to repulse a possible attack nor in deterring such an attack. Just as important are actions which lessen fear and tension between the countries concerned, reduce the likely gain or increase the likely loss to the invader from invasion. Such actions might include confidence-building measures, arms limitation and mutual progressive disarmament among potential antagonists to reduce tension between them. The normal gains of occupation can perhaps be denied by various forms of civil resistance, a possibility investigated in some detail by the Alternative Defence Commission (1983), while the governments of Holland and Sweden are actually funding research into the practicalities of this approach.

Inflicting loss on an invader is normally taken to imply military resistance or retaliation, an approach which has perhaps reached its apogee with the idea of mutually-assured-destruction-based nuclear deterrence. A much more positive (and probably less expensive) approach would be for countries to make certain benefits freely available to the international community, which would be forfeit if that country was attacked. For example, say that Britain were to spend half its defence budget on medical research, the results of which were freely distributed to other countries. Not only would another country be less likely to attack Britain because of direct forfeit of the medical benefits, but there would be great international pressure for Britain to be left alone. This is an example of a theme which recurs later: security through mutuality of interest.

Civil security

This essentially depends on the basic expectations of a country's population being met, or not being met for clearly understandable and justifiable reasons. As already noted, it is a security that is being increasingly undermined by economic and ecological decline. More broadly, and especially in poor countries, civil security is threatened by the failure of the conventional development process, in which so much has been invested. Instead of Northern industrial growth enriching 'underdeveloped' countries through trade and the transfer of technology in an atmosphere of North–South dialogue and co-operation, 'development' now often seems more like a process whereby Northern countries, through sheer economic power and domination of international financial markets and institutions, engineer the dependence of countries in the South (through indebtedness, soaring interest rates, deteriorating terms of trade, inappropriate trade) to effect a 'reverse transfer' of resources from South to North. The result, as is well known, is famine, food insecurity, rural impoverishment, burgeoning cities – all the familiar litany of problems that make civil insecurity a far greater threat to many governments than military aggression from outside.

The governments of Peru and Argentina have been among the first to react to this new perception – the latter cut its arms outlay from 4 per cent to 2 per cent of GNP between 1980 and 1984 for economic and social reasons, and Peru has promised action in a similar vein. As civil security becomes increasingly problematic in more countries, it must be hoped that more and more governments will reorder their priorities in this direction, rather than towards the inevitable alternative of greater internal repression.

China has shown that arms reductions in favour of civil benefits are not only the response of governments in desperate economic straits. Between 1971 and 1985 its arms spending fell from 17.4 per cent to (a still high) 7.5 per cent of GNP in 1985, with further big troop reductions announced in 1985 bringing it down to 4 per cent by 1988 (Brown *et al.* 1986, 1990).

Economic security

This will be dealt with in Chapters 5 and 6 with a discussion of the whole strategy of Another Development, launched by the Dag Hammarskjold Foundation (1975) and much developed since. Another Development is about people envisioning and controlling their development process in a way that makes the best use of their own resources to meet their fundamental needs, that is consonant with their own culture and traditions, and that conserves and regenerates their environment. Such a development would inevitably require fundamental social and political change in many countries. It could also quite revive the economies and cultures – and enhance the security – of countries in the South.

Ecological security

With regard to ecological security, Norman Myers has made a good case for this to be treated by many countries as the number one priority for any sort of national security (Myers 1986b). The term 'environmental bankruptcy' is gaining increasing currency (see for example Timberlake 1985) and more and more people are accurately fitting the characterisation 'ecological refugees' (those displaced from areas where environmental life-support systems have been seriously impaired or have ceased to function). Among key environmental factors, Myers lists soils, water, forests, grasslands, echoing a point made earlier: 'If a nation's environmental foundations are depleted, its economy will decline' (p.251).

As also noted earlier, the implications of environmental security are of more than purely national importance, because rivers, watersheds and climatic patterns, as well as human factors such as pollution and some nomadic migration, often transcend national boundaries. Here is another opportunity for nations to find mutuality of interest in the joint care of and responsibility for joint environmental resources in their own joint interest. This is no zero-sum security. The health of a shared resource enriches both communities and threatens neither. Its abuse damages both. Such shared resources provide countries with an opportunity, guided by modern scientific norms of sound ecological management, to make common cause where perhaps before none existed.

Through their pioneering work in the Rocky Mountain Institute in Colorado, USA, Amory and Hunter Lovins have effectively summarised the attributes of this new conception of security.

> A really secure society has certain necessities: water, food, shelter, a secure and affordable supply of energy. It embraces health, a healthy environment, a flexible and sustainable system of production, a legitimate system of self-government, a durable system of shared values. . . . But where can we get these things which so directly touch our lives and let us all feel safe? Most of all from the institutions nearest to us: from our own efforts, our families, our communities, our local governments. Real security comes less from central governments, dispensed from top down, than if we build it ourselves from the bottom up. But we cannot feel secure if we enjoy Life, Liberty and the Pursuit of Happiness while

others do not; for then at best we will feel uncomfortable, and we may even fear that others will come to take from us what they lack themselves. Thus we build real security above all when we strive to make our neighbours feel more secure, not less – whether on the scale of the village or the globe.

(Lovins 1986, p.16)

If the Palme Commission missed an opportunity to develop its seminal notion of an expanded definition of security, it also failed to give a convincing explanation for the causes of the nuclear and conventional arms race and why past efforts at disarmament worked or failed (mainly the latter), despite having the objective of such explanation explicitly in its terms of reference (p.187). The contradictions implicit in the arms race are too many and too obvious to be explicable purely on the basis of defunct notions of security or the dominance of military–industrial vested interests. These contradictions include the following points, some of which have already been made but which bear repetition:

- The pursuit of security by means which undermine that security.
- The amassing of nuclear arsenals which go far beyond that needed for the most 'mutually assured destruction' required for nuclear deterrence.
- A stated policy of nuclear deterrence but the development of weapons systems which, by speed, stealth, accuracy or defensive shield, tend systematically to undermine it through the perceived acquisition of counterforce capability, and consequent incentives for a first-strike (ICDSI 1982, p.105).
- The illogicality of seeking to defend a cherished ideology or values through the enactment of mutual suicide that could only result in the extinction of those values.

These contradictions point to deeper forces at work in the spiralling arms race, forces probably of great subconscious power which need to be exposed and demobilised if the arms race is to be reversed. Again, Amory and Hunter Lovins have succintly identified the deeper forces. 'The problem is war: and, underlying war, the legacy of tribalism, human aggression, injustice, power without purpose, the psychic premises of aeons of homocentric, patriarchal, imperial culture' (op. cit. pp.20–1).

These are the sorts of factors which James Thompson (1988) identified as the causes of 'social traps', one result of which he

21

perceived to be the arms race. One of the ways out of such traps is to 'change or supplement the reward–cost structure so that the trap becomes a trade-off' (p.43) and this is very much the approach President Gorbachev seems to have taken, unilaterally, with his disarmament initiatives. First, very much along the lines argued in this book, he perceived that the USSR's huge military expenditure was not in fact yielding security. As a corollary, it was clear that he could afford to reduce such expenditure very significantly without US or NATO reciprocation, even without materially affecting Soviet security (change in the reward–cost structure). Second, it was becoming ever more painfully obvious that arms spending was crippling the Soviet economy (the trade-off). From this viewpoint disarmament suddenly became an overriding rational necessity for the Soviets, even if it was unilateral, as evidenced by the fact that the arms agreements since then, signed or mooted, have in fact envisaged the destruction of more Soviet than American weapons. The Palme Commission quite failed to foresee the logic of such unilateralism.

A third example of omission or misplaced emphasis in the Palme Report is its almost exclusive focus on the role of governments in the disarmament process, despite a strong passing acknowledgement of the importance of the international peace movement:

> Recent years have witnessed the rebirth of popular movements determined to eliminate the danger of nuclear Armageddon. In Europe and, more recently, in North America, millions of men and women, mobilising impressive political strength, have demonstrated that fear of nuclear war remains an abiding concern. . . . That these movements have already strongly influenced events cannot be disputed; whether they can cause significant and permanent change in government policies remains to be demonstrated.
>
> (p.43)

Olof Palme, the Chairman of the Commission, goes further in his introduction to the Report:

> The tremendous popular and political awakening of the past two years has created a new public concerned with peace and security. . . . This popular insight is already a considerable political force and already has influenced events. It is very unlikely that disarmament will ever take place if it must wait for the initiatives of governments and experts. It will only come

about as the expression of the political will of people in many parts of the world. Its precondition is simply a constructive interplay between the people and those directly responsible for taking the momentous decisions about armaments and for conducting the complicated negotiations that must precede disarmament.

(pp.xii–xiii)

There is practically no space in the Palme Report given to the 'popular movements' for peace to which these quotations give so much political significance. Yet their vitality is a precondition of disarmament. A sample of their motivations, leaders and analysis is given in Chapter 3 of the present work.

THE BRANDT REPORTS

It is possible that, in retrospect, the two Brandt Reports will be viewed as the two concluding chapters of a view of economic development which dominated international relations for the three and a half decades following the Second World War. The characteristics of this development are amply evident in both reports:

- An almost exclusive preoccupation with nation states, the forces acting on them and the institutional and other relations between them.
- An overwhelming emphasis on industrialisation and economic (meaning production) growth as the substance of development.
- A predominant focus on financial institutions, instruments and indicators.

This is of course not to say that the Brandt Commission was unaware of the new dimensions to the development debate which had emerged during the 1970s, as shown by its several discussions of such topics as basic needs (BR1–ICID I 1980 pp.54ff), women (pp.59ff), environment (BR1 pp.113ff), human rights (BR1 p.10), people's organisations (BR2–Brandt Commission 1983 pp.80–1), 'informal' sector (BR1 pp.130–1), and participation (BR1 p.133). However, these citations practically exhaust the Brandt Reports' treatment of these issues which are now commonly perceived to be at the centre of any concern with economic progress. In both Reports' Summary of Recommendations these issues are either compressed into a couple of sentences or excluded altogether.

Most of the Reports' recommendations are to do with increasing financial flows from North to South, on more or less concessionary terms. This is especially true of the second Report, which was published after the extent of the Third World debt crisis had become apparent. Given that no such increased flows have transpired, the Brandt Reports have been a failure and it is worth enquiring why, in order better to assess whether other proposals might have fared better.

The first reason, sadly acknowledged by the Brandt Commission itself, is that the international climate for development co-operation in the 1980s was not encouraging. Both the US and UK were singularly unenthusiastic about development aid during these years, the Eastern bloc has been too concerned about its own economic problems, and the industrial countries that have been positively inclined, including Norway, Sweden, the Netherlands and Canada, have been unable to make the necessary global impact. The second reason, which is more to the point here, is that the flaws in the Brandt analysis made it unable either to change the unfavourable development climate or, more importantly, to indicate how the resources of progressive donors could be put to best use. These are the issues which will be briefly explored here, initially through the medium of the debt crisis, which is the most spectacular breakdown of the conventional development model.

During the 1970s many developing country governments borrowed heavily for development purposes from some or all of commercial banks, industrial country governments or multilateral institutions (such as the World Bank). The banks in particular were awash with petrodollars needing interest-bearing investments. These funds were spent, or allowed other funds to be spent, by developing country rulers in a wide variety of ways, including:

- potentially productive investments;
- unproductive investments;
- prestige projects;
- luxury imports;
- armaments;
- capital flight to private bank accounts abroad.

Unfortunately no detailed breakdown of Third World expenditure into these categories exists, not least because it would be highly embarrassing to some influential interests. However, some examples can be given to illustrate these points.

- Susan George (1988) has documented some of the more out-standing recent scandals of corruption and 'development'. Somoza, she claims, 'pocketed most of the international loans meant for the reconstruction of Managua and continued to steal from his country right up to the moment he was finally found out in 1979. . . . Nicaragua's debt is $4 billion, three-quarters of which was contracted under the Somoza regime.' (p.18). Marcos is estimated to have embezzled at least 15 per cent of the Philippines US$26 billion debt through such deals as creaming off US$80 million from a nuclear power station contract which had cost double that of a competing tender. Mobutu is thought to have stolen US$5 billion, the full extent of Zaire's debt (p.106). The Generals governing Bolivia spent hundreds of millions of dollars on, inter alia, the most expensive road per mile in the world, running from La Paz to the airport, and an oil refinery which has never operated above 30 per cent capacity.

Perhaps Bolivia was lucky to have seen any of its loans.

In 1986 Morgan Guaranty appraised capital flight from the big ten Latin American debtors (Brazil, Mexico, Venezuela, Peru, Colombia, Ecuador, Bolivia, Uruguay, Argentina and Chile) at 70% of all their new loans from 1983 to 1985. . . . In March 1985 a Mexico City newspaper published the names of 575 Mexicans who were all supposed to have at least $1 million deposited in foreign banks.

(ibid. p.20)

Brown et al. (1990) estimate Latin America's capital flight by late 1987 at US$250 billion. Venezuela's overseas holdings were put at US$58 billion compared to a total external debt of US$37 billion (p.144).

As for prestige projects, perhaps the prize for eccentricity must go to the President of Côte d'Ivoire who built a larger than life-size air-conditioned replica of St Peter's Basilica in Rome in his jungle. The building has nine acres of French stained glass and Italian marble and cost US$200 million, about 5 per cent of Côte d'Ivoire's GNP and the same as it spent in 1988 servicing its US$8 billion external debt. Alan Durning describes this new cathedral as 'an apt symbol for normal practices of development investment in much of the Third World' (Brown et al. 1990, p.143).

- Developing countries spent US$400 billion (1986 US$) of foreign exchange on weapons between 1960 and 1987 (Sivard 1989, p.19). In an earlier report, Sivard had noted:

> The investment in weapons increased faster than the ability of these developing economies to support it. Since 1960 their military expenditures in constant dollars have increased 5.3 times, their aggregate GNP 3.4 times and per capita GNP 2.0 times. The disparity shows in a mountain of debt and hundreds of millions of people still deprived of basic necessities of life.
>
> (Sivard 1986, p.12)

Sivard also shows in this report that in the early 1980s developing country expenditures on arms imports finally exceeded their economic aid from all sources, peaking in 1982 at more than US$30 billion.

- Many Third World elites have used development finance to pay for imported Western lifestyles rather than productive technology. Max-Neef *et al.* (1989) comment: 'According to estimates made by the economist Jacobo Schatan between 1978 and 1981 the amount of non-essential imports rose to $14 billion in Mexico, $10 billion in Brazil and $5 billion in Chile' (p.47). Dependence on such imports has several effects relative to the debt crisis:

  (a) Third World elites will wish to service their debts at practically any cost, in order to maintain the imports necessary for their Western lifestyles.

  (b) The money to service these debts and to continue the luxury imports, must increasingly come from annexing the subsistence resources of the poor or cutting back on their social provision, because of the absence of productive investments to generate the necessary foreign exchange.

The Brandt Reports almost completely ignore these failures to use foreign loans productively. They discuss these issues thus:

> Those who benefit most from the present distribution of wealth and economic power, whether in the North or the South, commonly fail to give the highest priority to their shared responsibility for improving the lot of the poorest in the world.
>
> (BR1 p.127)

and

> Waste and corruption, oppression and violence, are unfortunately to be found in many parts of the world. . . . We in the South and the North should frankly discuss abuses of power by elites, the outbursts of fanaticism, the misery of millions of refugees, or other violations of human rights which harm the cause of justice and solidarity, at home and abroad.
>
> (BR1 p.10)

Unfortunately neither Brandt Report does discuss these abuses of power, frankly or otherwise, which inevitably gives their call for massively increased financial transfers both an unreal and disturbing quality. For if previous financial transfers led to widespread misinvestment and misappropriation, what guarantees are there that precisely the same situation will not recur with new flows? If, as a distinguished left-leaning British academic has written, 'It is no use pretending that most Third World governments are anything except usually corrupt and ruled by elites of the worst sort' (Barratt Brown 1988, p.4); is it really realistic or constructive to ignore this fact and propose huge new transfers to precisely these elites?

In their analysis of what went wrong with development, the Brandt Reports concentrate on the failure of potentially productive investments to deliver the expected benefits. For this failure they allocate blame partly to developing country macro-economic policies but mainly to external forces beyond developing country control:

> It is true that some developing countries experiencing oil or other commodity booms in the 1970s embarked on ambitious surges of investment (often encouraged by the North as part of the recycling process) which later proved unsustainable. And the syndrome of overvalued exchange rates, lack of appropriate price and other production incentives (especially for exports and for agriculture), and lax credit policies with double- or even treble-digit inflation, can be found in certain countries. Such policies must be corrected. But the unfavourable external environment has exacerbated their problems beyond measure, and is forcing most countries, even many of the well-managed ones, into excessive retrenchment. 'The developing countries', as the World Bank President put it, 'are being battered by global economic forces outside their control.'
>
> (BR2 p.20)

These global economic forces include recession in the world economy, high interest rates, deteriorating terms of trade caused by falling commodity prices and protectionism against developing country manufactures.

It would clearly be foolish to deny the importance of these forces. And it would be wrong not to point out that these forces are more the product of deliberate policy in industrial countries than somehow beyond anyone's control. High interest rates were the result of burgeoning budget and trade deficits in the US. The fall in commodity prices is at least partially the result of the IMF's insistence that all developing countries pursue export-oriented strategies based on the same small number of primary goods, with inevitable consequent oversupply. Protectionism is obviously deliberate policy. The Brandt Reports are right to point out the inconsistencies between these policies and the West's rhetoric about free trade on the one hand, and about balanced budgets and sound economic management on the other. Moreover, in the years since the Brandt Reports, the impact of these factors on many developing countries has got very much worse, as some of the statistics quoted in Chapter 1 show. On the debt front alone, IMF figures show that for the fifteen most indebted countries, 1988 saw a reverse flow of funds from South to North (poor to rich) of US$24.5 billion, bringing to US$164 billion the net outflow since 1982 (quoted in Huhne 1989, p.89) while developing country debt levels from 1981 to 1987 have risen from US$748 billion to US$1,195 billion (ibid., p.96. Huhne gives his source for these figures as the IMF *World Economic Outlook*, October 1987 and April 1988).

These figures themselves undermine the heart of the Brandt argument for movement towards a new economic order based on mutual self-interest. The North should help the South to greater prosperity, his reasoning goes, not only on the grounds of equity, but because greater Southern prosperity is in the North's own interest because of greater demand for Northern imports. (Brandt also uses the argument of possible financial collapse due to a default on debt payments, but this possibility has receded to insignificance as banks have covered their exposure in the years since 1983).

The argument of Northern self-interest is very unconvincing. At present powerful Northern (creditor) institutions are reaping handsome returns on loans that never get any smaller; Northern industry and consumers are getting Southern commodities at rock bottom prices; Northern arms manufacturers are selling nervous Third World elites large quantities of weapons; and, perhaps most important, Third World countries are being effectively tied in to Western models of development, with Western countries guaranteed a permanent technological lead. The old economic order is doing industrial countries very nicely economically and politically. Third World resources are as cheaply available as in colonial days without the costs of foreign administrations. A prosperous Third World would assuredly revive the talk of the 1970s of 'self-reliance' and 'culturally-appropriate development' which would not at all be in perceived Northern self-interests.

Thus the Brandt Reports, for all their technical insight and expertise and avowed commitment to Third World development, remain hopelessly flawed:

- They fail to identify the major cause of development failure contributed by corrupt and expropriatory Third World governments and inappropriate development models. This means that even if their recommendations were adopted (i.e. greater financial flows) they would be more likely to lead to a repeat of development failure than development.
- They postulate a mutual self-interest in reform where none in fact exists. This means that their recommendations are most unlikely to be implemented.
- They fail to identify with anything like enough emphasis the real reforming agenda which could break the development deadlock: strategies of participation, community organisation, democratisation, feminisation and environmental conservation.

It is the reforming agenda, and the actions taken to implement it, that will be the subject of Chapters 5 and 6. With hindsight it is not surprising that the Brandt Reports were unable to get to grips with it for this agenda comes authentically from the grassroots, from the bottom up. The Brandt Commission, on the other hand, was composed of top people, thinking top down, as such people normally do. The problem with their top-down recommendations was that other top people, who would have had to have implemented them, were and are doing very well out of the status quo. The recommendations simply died on the page.

THE BRUNDTLAND REPORT

The Brundtland Report of 1987 took up many of the themes of the Brandt Report but sought to relate these themes of development explicitly to environmental deterioration, advocating in solution a process called sustainable development. Its analysis broadly concluded that rich, mainly industrial country people are destroying the environment through high consumption lifestyles, and the economic activity needed to generate them, which does not take account of its environmental impact; and that poor, mainly Third World people are destroying the environment in their struggle to stay alive.

The Brundtland Report, the largest of the independent UN Commission reports surveyed here, examines the implications of these conclusions in very great detail, across the whole range of relevant issues: population and human resources, food security, species and ecosystems, energy, industry and urbanisation. It also explores the issues of the global commons, security and peace, and institutional reform. Many of its concrete recommendations would clearly contribute to a diminution of the environmental problem. Moreover, the Report is far more sensitive to the new 'reforming agenda' which the Brandt Reports had dealt with in such summary fashion. It places far more emphasis on voluntary organisations, whose evidence to the Commission is frequently quoted verbatim in the report (the Commission had innovatively held public hearings round the world specifically to gather such evidence); its criteria for sustainable development specifically include:

> a political system that secures effective citizen participation in decision-making.
>
> (WCED 1987, p.65)

it advocates help for smaller rural producers:

> Smallholders, including – indeed especially – women, must be given preference when allocating scarce resources, staff and credit. Small farmers must also be more involved in formulating agricultural policies.
>
> (p.143)

It is also recommends 'decentralising the management of resources upon which local communities depend, and giving these communities an effective say over the use of these resources' (p.63).

However, the Brundtland Report makes one serious political error. Notwithstanding that its analysis reveals two very different causes of environmental destruction – over-consumption by rich people and countries, poverty in poor countries – it recommends a single over-arching solution to the problem, which occupied the key promotional position when the Report was launched. This solution was further economic growth.

> What is needed now is a new era of economic growth – growth that is forceful and at the same time socially and environmentally sustainable.
>
> (p.xii)

and

> It is essential that global economic growth be revitalised. In practical terms, this means more rapid economic growth in both industrial and developing countries, freer market access for the products of developing countries, lower interest rates, greater technology transfer and significantly larger capital flows, both concessional and commercial.
>
> (p.89)

This is pure, conventional developmentalism of the Brandt variety, but is more dangerous related to Brundtland's concern with sustainability. The problem with calling for more economic growth in this way is that nowhere in the Brundtland Report is there a clear statement of how 'sustainable economic growth' can be recognised and distinguished from the patently unsustainable variety which is all the industrial world has so far known and which was largely responsible, by the Commission's own analysis, for the environmental destruction which led to it being convened. It is very likely, therefore, that the call for more economic growth will be interpreted as an endorsement of business-as-usual, economic growth having been for some decades the top policy priority, without the thorough policy overhaul which the Brundtland Report elsewhere advocates. Such an interpretation was all too evident in the complacent response of the UK government to the Report, indicated in the following quotations:

> The central message of the Report is the need for a new era of economic growth and the integration of environment and development. The philosophy of zero growth is rejected.

Growth is seen as both necessary and possible provided that it is sustainable.

(DOE 1988, p.11)

One of the major objectives is the revival of growth in both industrial and developing countries. The rejection of zero growth is welcome.

(ibid., p.13)

Rather than simply calling for more economic growth, the Brundtland Report should have boldly followed through the obvious conclusions of its excellent analysis, giving different recommendations according to the level of industrialisation:

Industrial countries and Third World industrial sectors

These should do a root and branch, sector by sector analysis of the environmental impacts of their economic activity, in terms of resources (renewable and exhaustible) used, wastes emitted and biosystems affected, and take the necessary steps to bring these impacts within defined sustainability criteria. Such steps might or might not lead to economic growth. They would certainly bring about a radical restructuring of production and consumption in which polluting or depleting activities were either made more efficient or curtailed, while activities without such environmental impacts enjoyed a relative price advantage. Organic agriculture would benefit over the intensive chemical variety. Cyclists and those using public transport would gain over private motorists. Investment in energy conservation and efficiency would yield a greater return relative to energy use. This could not be a Pareto no-lose process, for some people would inevitably be worse off. Some prices would rise and some activities diminish. This would reduce economic growth and there is no guarantee that substituting activities or technologies would fully compensate for this. To recommend economic growth in this context is simply to miss the point.

Developing countries

These should continue to strive for greater production and productivity, but only on the basis of the increased sustainable production of biomass from a regenerated environmental base. As Robert

Chambers has observed (e.g. Chambers 1988 pp.17ff), many degraded Third World environments can become sustainably productive again, but only with appropriate investment, and above all, the full participation and control of the intended beneficiaries of the increased production. This would have the further beneficial effect of stemming the flow of people from the countryside to the towns and so contributing to the solution of another major social development problem. It is evident that the sort of development strategy which concentrates almost exclusively on the participatory rural regeneration of ecosystems to yield sustainable livelihoods is about as different a strategy from the current priorities of the World Bank and developing country governments as it is possible to imagine. Yet such a strategy is both implicit in the Brundtland Report in its discussion of developing countries and, as some of the quotations given have shown, comes close to being explicit on occasions.

As far as Third World industries are concerned, a conversion to sustainability will only be achieved in the context of a very large transfer of training and technology on concessionary terms from industrial countries. It is simply inconceivable that governments of populations, a large proportion of which are living in absolute poverty, are going to be able successfully to demand sacrifices from both their poor and middle classes for a sustainability the benefits of which may not be apparent for two or three decades, unless they are perceived to be receiving real backing for the future from the rich world, while it simultaneously puts its own house in order. It is worth remembering in this context that no action for sustainability in the West will be effective if China and India, with their 36 per cent of the world's population, proceed with their current environmentally ruinous industrialisation programmes. The governments of such countries have simply got to be persuaded to respect the global environment and this persuasion will be expensive. The necessary transfer from rich to poor countries will, of course, have further negative implications for economic growth in rich countries.

There are as yet no signs that this dual message has been either heard or understood by the big multilateral development institutions or by any of the world's governments.

INTERDEPENDENCE AND SUSTAINABLE DEVELOPMENT

The Brundtland Commission's 'Overview' which opens its Report is revealingly entitled 'From One Earth to One World', reflecting two important facts. The first is that, although the human race has always lived on a single Earth, the time has now passed when we can pretend we do not. The growth of population and technological change have demolished the buffers of time and space which used to a large extent to separate different societies and set up an intricate web of global interdependencies in every sphere: ecological, economic, social and cultural. However, the second point is that notwithstanding these interdependencies there still exist many 'worlds' as experienced by different human populations, both between North and South and within individual countries in those imprecise categories.

Calling for a transition to 'one world' as the Brundtland Report does, implies a diminution of this diversity of experience. Indeed, Wolfgang Sachs (1989) has argued tellingly that forced cultural convergence has always been a fundamental component of the 'development' project established after the Second World War. The very notion of development comes ready made with its own hierarchy and hegemony expressed by such terms as 'advanced', 'developed', 'developing' and 'underdeveloped', all of which are in common use. Once 'development' became the principal objective of practically every country in the world, a global process of homogenisation in emulation of the world's most 'developed' country, the United States, was inevitable. The process is made possible by the web of interdependencies that now exist. These interdependencies are often used as an argument for greater caring and sharing in the world economy, but this overlooks the fact that relations of interdependence can as easily be hegemonistic and exploitative, as in colonialism or master/slave relationships, as based on co-operative mutuality. Whether or not they are so depends on the institutions of interdependence through which the relationships between countries are expressed.

Institutions of interdependence: trade

The ideology of free market capitalism, and the economic theory behind it, holds that trade promotes development because by definition it is good for all parties, being based on voluntaristic mutual gain. I have elsewhere (Ekins 1989) contested the validity of that view on the grounds that:

1. Differences in power between the trading parties may result in all the gains from trade going to the more powerful partner (exploitative trade type I);
2. Differences in power within one of the trading parties may result in the gains from trade being unjustly distributed between constituent groups (exploitative trade type II) as well as contributing to the oppression of less powerful groups.
3. Trade can lead to dependency relationships (e.g. for food, capital goods, weapons, luxury imports) which effectively remove the freedom *not* to trade. Trading decisions can also be greatly constrained by such external factors as indebtedness and being locked into the export of primary commodities by the batteries of Northern protective measures which limit the extent of Southern diversification into secondary activities.
4. Trade, through dependency, can lead to vulnerability, insecurity and loss of autonomy.

There are other reasons why the classical model of trading relationships established on the basis of mutual gain from comparative advantage, and mediated by prices formed by some interaction between cost of production and supply and demand, now bears little relationship to the global marketplace.

Firstly, as Daly and Cobb (1990, pp.209ff) have convincingly argued, the very notion of comparative advantage has been rendered inoperative by the relatively new situation of high capital mobility, which makes *absolute* advantage the determining factor as to where production is to be located.

Secondly, such a model always underestimated the role of policy and power, political, military or economic, in setting prices. Industrial country agricultural subsidies and the consequent dumping of surpluses depress world food prices while keeping them high in protected Northern markets; protection of Northern markets in, say, textiles, diminishes the effect of the price mechanism; Northern-instituted IMF structural adjustment programmes meanwhile have brought about increases in supply of a limited and relatively demand-inelastic number of commodities by Southern countries, resulting in a fall in their price relative to Northern exports (exploitative trade type I).

Thirdly, the price mechanism is increasingly being bypassed altogether, both by the growth of countertrade, whereby goods are exchanged directly at an implicit price rather than sold on world markets, and by the growth of intra-corporate trade, which has now been estimated to account for 30 per cent of all trade. In reality, of course, it is not 'trade' at all, for despite involving a currency transaction, it does not involve exchange, being more in the nature of a bookkeeping operation within a single organisation.

The growth of intra-corporate trade, and indeed of the transnational corporation itself, provides an interesting countertrend to the spread of the global market. The largest corporations now have a higher turnover than most countries. In terms of structure they resemble nothing so much as the former command-and-control economies of Eastern Europe and the USSR which have now acquired such a bad name: within them resources are allocated through hierarchical dictat rather than by market forces. These giant corporations therefore represent great islands of central planning within the global market. Moreover, because of their economic power, they are able greatly to broaden the range of price-setting possibilities, including transfer-pricing, predatory pricing, oligopolistic pricing, etc.

All these factors together mean that the global trading system, far from being a mutually beneficial voluntaristic system of exchange, has become a means of coercion, employed jointly by powerful institutions in the First World and their client elites in the Third, by which to force Third World resources into the global market on terms highly unfavourable to the vendor. Needless to say, this process also undermines traditional culture and social structures conducive to self-reliance, so that it generates a dependency on Western imports and Western lifestyles which becomes increasingly difficult to overcome.

Institutions of interdependence: aid

It would seem that trade, while undeniably an instrument of interdependence, is not proving conducive to co-operative mutuality. Unfortunately, precisely the same can be said of aid, despite the fact that its name has connotations of help and altruism.

In his study of aid and development, Hancock (1989) showed that the US$60 billion in aid money which annually flows between governments overwhelmingly serves the following purposes:

1. Promotion of the domestic economic or foreign policy interests of the donor country.
2. Personal enrichment, gratification or support of powerful individuals in the recipient country.
3. Maintenance of the luxurious lifestyles of development professionals, especially in the United Nations and multilateral development organisations.
4. Multiplication of bureaucracy with associated waste, inefficiency and abrogation of responsibility.

To these objectives which have self-evidently nothing to do with serving the interest of the poor in recipient countries can be traced most of the development disasters, documented by Hancock, which have afflicted Third World countries:

> Roads that end in rivers and then continue blithely onward on the other side, silos without power supplies, highly sophisticated equipment that no-one can use, installed in remote places, aquaculture projects producing fish at $4,000 per kilo for consumption by African peasants who do not even earn $400 per year, dams that dispossess thousands and spread fatal water-borne diseases, resettlement schemes that make the migrants poorer than they were before they left home, that destroy the environment and obliterate tribal peoples – such blunders are *not* quaint exceptions to some benign and general rule of development. On the contrary, they *are* the rule.
>
> (Hancock 1989, p.148)

In fact it is somewhat surprising that a relationship between governments of North and South should ever have been expected to benefit the Southern poor, because the majority of Southern governments are neither representative of nor sensitive to the interests of their poor. As already discussed, sixty-four developing countries were under military control in 1988, and forty-three countries were reported to the UN Special Rapporteur on Torture for torturing their citizens. Yet many of these governments of these countries are precisely those to which it is often proposed that ever greater transfers of 'aid' should be effected.

Institutions of interdependence: debt

The phenomenon of Third World debt, as discussed earlier, also well illustrates the kind of exploitative structures through which Northern interests predominate in the world economic system.

Western banks and governments lent the huge sums involved to the people running Third World governments in the full knowledge that these people were, in the main, quite unrepresentative of their countries' populations as a whole, and that most of the money would be spent in a variety of unproductive ways. They thus ignored standard principles of banking caution and responsibility.

Since the loans have predictably failed to perform, threatening to cut off further flows to the Third World elites who had contracted them, these elites have sought to force the poor of their countries to pay for the debt-service, either by sequestering their natural resources, especially land, for redirection to the export sector, or by cutting social security arrangements. To implement these measures the elites themselves have been either encouraged or coerced by the structural adjustment policies of the Western-dominated IMF and World Bank. It is an astonishing situation, a classic example of exploitation by interdependence, which contradicts both banking and natural justice. Lenders and borrowers have ganged up on a third party, who was in no way involved in their transaction, and who happens to be far poorer than either of them, to extract the resouces so that the borrower's debt-obligations may be repaid to the lender. Yet, as with aid, well-meaning people often propose that still more money should be lent into this deeply exploitative relationship.

Institutions of interdependence: people's organisations

The institutions of interdependence so far discussed not only exploit the poor, they are fundamental contributors to unsustainability, ensuring that Southern environments are systematically ransacked of resources for the overconsumption of the rich, resulting in the marginalisation of the poor, which Chapter 7 shows to be the two principal reasons for environmental destruction. But there is another set of institutions active in the development process, people's organisations, which are the antithesis of the conventional aid/trade/debt links in the global economy.

For Bertrand Schneider these organisations are the only entities capable of tackling the three requisite tasks for improving the quality of life of the rural poor, whom he calls 'the real challenge of the development process' (Schneider 1988, p.229):

1. The removal of the 'factors of impoverishment afflicting rural populations'. Such factors essentially comprise misguided policies, corruption or outright repression by their own governments, or destructive 'development aid' by others.
2. The definition by villagers of their basic needs and how they can best be met.
3. The development by villagers, with appropriate external assistance, of the necessary factors of production to meet these needs in the desired fashion (p.223).

The growing involvement of people's organisations in development work Schneider calls 'the barefoot revolution'. In the South their approach to development tends to combine poverty alleviation with environmental concern because it is precisely through the regeneration of environmental resources that they hope to alleviate their poverty. The ecosystem is the most important 'factor of production' in meeting their needs. In the North non-governmental organisations are increasingly linking with their colleagues in the South in an effort to bypass or reform the destructive institutions of interdependence. Groups in both North and South have perceived far more clearly than their governments the great unarticulated imperatives of the Brundtland analysis:

- The priority of the reform of Western production and consumption rather than economic growth;
- The regeneration of biomass worldwide but especially in the South;
- The need for Southern industrial transformation through technology transfer and a switch to the sustainable use of resources.

It is the non-governmental mobilisation to meet the joint tasks and imperatives of development and environment that will be described in Chapters 5 to 7.

3

PEACE THROUGH PUBLIC PRESSURE AND REAL SECURITY

THE POWER OF PUBLIC OPINION

The years since the Palme Commission report have seen great changes in international relations, principally between the First and Second Worlds. Since the advent of Mr Gorbachev, the threat of East–West conflict has receded to its lowest point since the end of the Second World War. As importantly, changes in Eastern Europe, which are far less driven by a single individual and are therefore less vulnerable to reaction, have irreversibly altered the post-war division of Europe.

These developments are obviously of immense significance and they bring the prospect both of significant superpower nuclear disarmament and a diminution of the regional conflicts like those in Angola and Afghanistan in which superpower proxies have fought each other with their patrons' weapons at increased cost to the local people. But though an improvement in East–West relations was probably a necessary condition for a new regime of global security, it is by no means sufficient. Military spending remains very high and the proliferation of weapons of mass destruction proceeds apace. Both Israel and South Africa are reliably thought to have entered the nuclear club, with an Islamic bomb in Pakistan, and others in India, Latin America and elsewhere, probably in the making. Many other countries are now turning their attention to chemical weapons, after their 'successful' use by Iraq in the Iran–Iraq war, which drew such a strangely muted response from international opinion. With the potential for indigenous regional conflicts remaining as high as ever, and with terrorism also common in many countries, the spread of these weapons is extremely dangerous.

While the complexities of the situation are such that one will probably never be able to specify direct lines of cause and effect for the improvement in East–West relations, it is possible to make some general statements with a high probability of accuracy. The first is that whatever the USSR's economic and other internal imperatives, it is most unlikely that Mr Gorbachev could have proceeded with his disarmament initiatives had the West seemed to Soviet eyes to be likely to attack the Eastern bloc. The response of Western public opinion to the Cruise/Pershing decision in 1979, and the large demonstrations for peace which occurred in Western capitals during the early 1980s must surely have reassured the USSR that, despite the warlike rhetoric of President Reagan, Western people would simply not tolerate Western aggression against the Warsaw Pact. Western public opinion has been similarly influential in causing Western leaders to do business with Mr Gorbachev when the early signs were that NATO was distinctly unwelcoming to his proposals and, in West Germany, it was directly successful against immense NATO pressure, in postponing the 'modernisation' of tactical nuclear weapons in Europe, which the leaders of several other Western countries were determined to enact. Even Mrs Thatcher was forced to recognise that NATO's nuclear weapons in Germany, far from being modernised, will have to be phased out (Pick 1990, p.1).

It is also necessary to pay tribute to the vision of the Western peace movement which saw past and through the Cold War even at the height of its freeze in the early 1980s. Such vision is perhaps best expressed in the foundation in 1980 and subsequent work of European Nuclear Disarmament (END) and in the speeches and writings of END's co-founder, the British historian E.P. Thompson.

Thompson had already been the prime mover in the most important single publication of the British peace movement, *Protest and Survive* (Thompson and Smith 1980), when he helped to draft the founding END Appeal with its objective 'to free Europe from confrontation, to enforce detente between the United States and the Soviet Union, and ultimately, to dissolve both great power alliances'. At a time when the 'evil empire' rhetoric was at its height, and NATO and the Warsaw Pact had never seemed more entrenched, Thompson was writing and giving speeches like 'Beyond the Cold War' (Thompson 1982). He always saw that:

> If this adversary posture were to be modified, if it were to be undermined by new ideas and movements on both sides, then not only the weapons but the launch-pad for them would be taken away ... We must go behind the missiles to the Cold

War itself. We must begin to put Europe back into one piece. . . . Something remarkable is stirring in this continent today; movements which commenced in fear and which are now taking on the shape of hope; movements which cannot yet, with clarity, name their own demands. . . . These voices signal that the whole 35 year old era of the Cold War could be coming to an end.

(Thompson 1982, pp.2, 25, 27)

This was in 1982.

END also nurtured and supported the East European independent peace and human rights groups. Its Appeal in 1980 was signed by Vaclav Havel (now President of Czechoslovakia), Jiri Dienstbier (now Czechoslovakia's Foreign Minister), Andrea Hegedus (now Hungary's Prime Minister), the two co-founders of East Germany's Green Party and the General Secretary of East Germany's Social Democratic Party. END networked ceaselessly with the fragile persecuted groups these people then led, it published their writings, it held meetings and conferences, it campaigned for them in the Western press while insisting that their repression should not be used as an excuse for Western militarism for their 'freedom'. This initiative must surely have contributed greatly to the momentous events of 1989 in Eastern Europe when these groups blossomed into the authentic expression of their fellow-countryfolk's will. The prescient Thompson had seen the writing on the wall a full two years earlier when he said in a speech to END:

The world that we seek to realise is already here. It needs only the courage to throw off the shadows of the past, to show itself, to believe in its own existence. The ideologists offer to us a present which is already dead and which is dragging out a posthumous existence on the life-support system of militarism. But human consciousness has changed, and we have helped to change it. . . . Everywhere, on both sides, people know that a profound change is to come. It is our business to disclose the present as it really is.

(Thompson 1987 pp.5, 6)

Following 1989's East European revolutions, END through a new organisation, European Dialogue, helped to set up the Helsinki Citizens Assembly to parallel the governmental Helsinki Process, the Conference on Security and Cooperation in Europe (CSCE). The

Assembly was founded in Prague in October 1990 when 700 people from the thirty-five CSCE countries gathered as a kind of 'People's Parliament from below' working for 'the democratic integration of Europe and for the peaceful and constructive resolution of conflict' (European Dialogue 1990).

Thus whatever the predisposition of government, an alert, active and informed public opinion would seem to be a precondition for positive change in the security field as in many other areas of public policy. It was the mobilisation of such public opinion that was the objective of the peace organisations which mushroomed across Western Europe and North America following the Cruise/Pershing decision of 1979. Tens and sometimes hundreds of thousands of people marched through the major cities of these regions inspired by a new analysis of militarism purveyed by a new generation of leader-activists who were more often than not women. This analysis is well exemplified by the life and work of three women – Petra Kelly, Helen Caldicott and Astrid Einarsson – who are briefly profiled here.

Petra K. Kelly (Germany)

Petra K. Kelly was born in 1947 and studied in the Netherlands, Germany and the US before joining the Economic and Social Committee of the European Community in 1971, working on social, environmental and women's policies. During the 1970s she became increasingly active in the European women's, peace, and anti-nuclear movements, having also been a protester in the US against nuclear weapons and the Vietnam War. In 1979 she helped to found the West German Green Party, die Grünen, and was one of twenty-seven Greens elected to the Federal Parliament, the Bundestag, in 1983, where she was elected one of the Green parliamentary speakers. She remained in the Bundestag until 1990.

> Splitting the atom; uncontrollable emission of radioactive toxins; the insanity of the nuclear, bacteriological and chemical weapons build-up; unrestrained economic growth spreading commercialisation to every aspect of our lives; overconsumption of goods and raw materials; the erosion of the individual's right to free speech; anti-human architecture, transport, technology and food production; increasing indifference and irresponsibility on the political front – these are the conditions of modern industrial society.
>
> (Kelly 1984, pp.92–3)

To counter these conditions Kelly advocated four kinds of action:

> *Legitimate action* ... (including) readers' letters, signing petitions, distributing leaflets, demonstrating and knocking on doors. ... *Symbolic action* ... in vigils, silent marches, fasts, as well as more light-hearted events. ... *Non-cooperation* ... such as strikes, boycotts, conscientious objection and non-acceptance of state honours ... (and) *civil disobedience* – open infringement of the law on grounds of conscience.
>
> (ibid. pp.31–2)

For Kelly these actions were relevant and necessary across the range of issues on which she concentrated her political work and speech-making: peace and non-violence, ecology, feminism and human rights. She constantly sought to tie these themes together into a single problematique.

> We are watching cowboy economics and cowboy threats; we are watching an industrial order with its expansionist, machismo, militaristic, and patriarchal nation states. Confronting the system of machismo is a trend towards new-age politics, a trend towards eco-feminism. We try to make others aware of such basic principles as the value of all human beings and the right to satisfaction of basic human needs within ecological tolerances of land, sea, air and forests. All these principles apply, with equal emphasis, to future generations of humans and their biospheric life support systems, and thus include respect for all other life forms and the Earth itself. ... What we have in common in the ecological, feminist and peace movements is not small. ... The arms race, I believe, is insane, but an inevitable outcome of science in a world where men wage war against feminine values, women and nature.
>
> (ibid. pp.38–9)

Helen Caldicott (Australia)

While Kelly was helping to lay the foundations of a West German peace movement which, in 1989, made it impossible for a conservative West German Chancellor to modernise tactical nuclear weapons, against the full weight of NATO pressure, Dr Helen Caldicott was mobilising thousands along similar lines on the other side of the Atlantic.

Caldicott had cut her campaigning teeth in 1971–2 when, in her native Australia, she had protested successfully against French atmospheric nuclear tests in the Pacific, driving them underground; and blocked uranium mining in Australia for seven years by mobilising the Australian trade union movement against it. In 1977, as a medical doctor, she took a teaching post at Harvard Medical School in the US, but very soon became deeply involved in the nuclear issue. In 1978 she founded and became President of Physicians for Social Responsibility, a post she held for five years, during which she took the organisation's membership to 30,000, transforming it into one of the most effective peace organisations in the US. In 1980 she founded Women's Action for Nuclear Disarmament which concentrates on lobbying politicians both in Washington and, by supporting its local members and groups, in their constituencies. During this period Caldicott was called 'America's most effective one-woman recruiting agent for the anti-nuclear cause', not only informing people through many speeches, press interviews and TV appearances, vividly conveying the physical and medical horrors of nuclear war, but effectively convincing them to become active on the issue. Caldicott's analysis, as expressed in her book *Missile Envy: the Arms Race and Nuclear War* (1985), is very similar to Kelly's. Firstly, there is the common conviction that the arms race is a male-dominated and male-inspired aberration:

The hideous weapons of mass genocide may be symptoms of several male emotions, reflecting inadequate sexuality, a need continually to prove virility, and a primitive fascination with killing. . . . Women have a very important role to play in the world today. They must rapidly develop their own power so that they can move out into local, national and international affairs. . . . I don't mean that in doing this women should abrogate their positive feminine principle of nurturing, loving and caring. I mean they should tenaciously preserve these values but also learn to find and use their incredible power. The positive feminine principle must become the guiding moral principle in world politics.

(p.241)

Like Kelly, too, Caldicott explicitly linked peace and ecological issues, advocating 'conversion from a war economy to a peace economy'.

The world urgently needs adequate production and equitable distribution of food. . . . Reforestation of many areas of the world is a mandatory priority. . . . The riches of the sea must be equitably distributed among all nations on earth. . . . All the world's natural resources must be enhanced and used for the family of man and not hoarded and wasted on the production of weapons. . . . The air and water of America and large parts of the world are fast becoming irretrievably polluted with carcinogenic and mutagenic poisons. . . . Most have never been adequately tested for carcinogenicity and most are released to the environment, often illegally. Many of these chemicals are by-products of industries that produce plastic throwaway materials we don't need.

(pp.306, 307)

Caldicott was writing in what many people considered to be the darkest and most dangerous period of the 'new Cold War', the years of the 1980s before the emergence of President Gorbachev, when Western arms-spending and anti-Soviet rhetoric both reached new heights. She prophetically advocated a six-point plan to move towards disarmament.

To stop the nuclear arms race effectively and to move rapidly to bilateral nuclear disarmament, several simultaneous steps need to be taken: (1) a nuclear weapons freeze; (2) a nonintervention treaty signed by the superpowers, which will end military intervention in developing countries and in superpower satellites; (3) a reduction in the huge conventional forces of NATO and the Warsaw Pact and in the conventional forces of Japan and China; and (4) an end to the massive international arms trade to Third World countries, which would be destabilizing in itself, even if the superpowers ceased their interventionist tactics; (5) an end to innovation and development of conventional weapons – developments in fighter planes, tanks, ships, and missiles create instability and uncertainty about the future; and (6) at the same time, a move toward rapid bilateral nuclear disarmament.

(p.165)

The years since 1985 have indeed seen simultaneous progress on most of these points. Bilateral nuclear disarmament (6) has

46

proceeded through the INF agreement of 1987 and the Strategic Arms Reduction (START) talks, currently aiming to cut superpower nuclear arsenals by 50 per cent. Superpower intervention in developing countries (2) has been reduced by the Soviet withdrawal from Afghanistan and the termination by the US of military aid for the Nicaraguan Contras. Peace treaties involving the withdrawal of foreign troops have been concluded in Angola and Namibia. Talks on the reduction of conventional arms in Europe (3) envisage deep cuts on both sides. Although a nuclear freeze has not been accomplished, the modernisation of tactical nuclear weapons in Europe has been stalled and the US Strategic Defense Initiative (Star Wars) has been curtailed in response to Soviet peacemaking initiatives.

Of course, none of these developments is irreversible, nor have they yet made any real impact on arms budgets, nor have there been any moves to reduce the international trade in weapons. The peace process thus remains partial and fragile, with the powerful vested interests of the military–industrial complex still commanding enormous influence. However, despite continuing uncertainties, the events of 1989 in the USSR and Eastern Europe have done much to remove the image of 'the enemy' which has justified the arms race in the past. Provided that the perceived need for disarmament does not recede with the perception of the Communist threat, and public awareness of the gross wastage of resources that is entailed in arms budgets at current levels continues to grow, prospects for a significant reduction in those budgets must now be better than at any time in the immediate past.

Astrid Einarsson (Sweden)

It was to bring about the *realisation* of these prospects that Astrid Einarsson had the idea of the Great Peace Journey. While Kelly and Caldicott were mobilising in Europe and the US respectively, Einarsson, a Swedish high school teacher and long-time activist on a wide variety of issues, including peace, through the Women's International League of Peace and Freedom, has resolutely campaigned at the global level. In 1982 she participated in the peace march to Moscow and it was on this journey that the idea of the Great Peace Journey was born.

The basis of the Great Peace Journey is five questions on peace issues, based on the UN Charter, which were to be put by special delegations to all the member governments of the UN. The questions were:

1. Are you willing to initiate national legislation which guarantees that your country's defence forces, including 'military advisers', do not leave your territory for military purposes (other than in United Nations peacekeeping forces) – if all other Members of the United Nations undertake to do the same?
2. Are you willing to take steps to ensure that the development, possession, storage and employment of mass-destruction weapons including nuclear weapons, which threaten to destroy the very conditions necessary for life on this earth, are forbidden in your country – if all other Members of the United Nations undertake to do the same?
3. Are you willing to take steps to prevent your country from allowing the supply of military equipment and weapons technology to other countries – if all other Members of the United Nations undertake to do the same?
4. Are you willing to work for a distribution of the earth's resources so that the fundamental necessities of human life, such as clean water, food, elementary health care and education are available to all people throughout the world?
5. Are you willing to ensure that any conflicts, in which your country may be involved in the future, will be settled by peaceful means of the kind specified in Article 33 of the United Nations Charter, and not by the use or threat of force?

(GPJ 1987, p.3)

By May 1989 over 100 countries had been visited and the governments of ninety-one countries had answered YES to all five questions, including Brazil, China, India, Israel, Pakistan and the USSR, but excluding France, the UK, US and West Germany. More YES signatories are still being sought and the global network of thousands of activists which the visits have helped create are seeking ways to persuade governments to implement their pledges. No government has yet been found to answer NO to any of the questions.

In September 1988 the Great Peace Journey organised the First Global Popular Summit at the UN HQ in New York, using the slogan 'We the Peoples of the United Nations *Still* Determined to Save Succeeding Generations from the Scourge of War'. One outcome of the Journey has been the beginning of an effort to establish Peace Zones in various regions of the world with YES governments (like the Nordic countries) as 'areas where the fundamental ideas of the United Nations about peaceful coexistence and global solidarity will be realised in practice' (Einarsson 1989, p.1). These Zones will be immediately recognised as examples of the Palme Commission's 'Common Security' . The process that is seeking to bring them about is a striking vindication of Palme's expressed belief some years earlier, quoted in Chapter 2, that governments would only move towards disarmament if pushed by popular pressure.

Yesh Gvul (Israel)

All the sources so far quoted in this study have stressed that regional tension and wars remain a significant source of global instability and human misery. No region has been more violent in recent years than the Middle East and it is there, in Israel, that a new peace movement has emerged of great importance both for Israel and further afield. The movement is called Yesh Gvul (literally 'there is a limit') and it consists of Israeli soldiers who are refusing to fight or serve with the Israeli Defence Forces in the Occupied Territories.

Yesh Gvul was born in the aftermath of the Israeli invasion of Lebanon, when Israel's flagrant aggression finally overcame inculcated habits of duty and obedience and soldiers refused to be posted there. By the end of the Lebanon campaign, at least 160 'refuseniks' had been jailed, with commanding officers failing to prosecute many more. Such insubordination in the Israeli army was unknown and had a profound impact on public opinion. Yesh Gvul claims that 'Fear of a further dramatic rise in the number of refuseniks was a prime factor in the Israeli government's ultimate decision to call off the campaign' (private communication from Yesh Gvul spokesman, 18 April 1989).

Yesh Gvul declined following the Lebanon withdrawal but revived with the *intifada*. By April 1989 sixty-three soldiers who refused to serve in the occupied territories had been jailed.

Dudu Palma, a paratrooper with extensive combat experience who has been jailed three times, expresses the refusenik's position thus:

> Feeling responsible for the future of Israeli democracy, I can no longer be party to anti-democractic acts verging upon war crimes. It is incredible that a people which so recently savoured its own political independence, should so lightly deny it to members of another people. By this step, I believe I am defending our fragile democracy which is being swept to the precipice by the rising tide of nationalism and Khumeinist fundamentalism.
>
> (quoted in Yesh Gvul undated, p.1)

Yesh Gvul has developed sophisticated support services for its members when they are under pressure. One briefing document, published in the style of an army 'Service Paybook', stresses the illegality in international law of Israel's treatment of Palestinians. Another offers help, advice and support to refuseniks, would-be refuseniks and their families while refuseniks are in prison. Another major Yesh Gvul activity is the publication in major national newspapers of a letter explaining their refusal. Over 1,000 reservists have already signed the letter, while a junior Yesh Gvul has also started consciousness-raising among high-school students.

Yesh Gvul's action of refusal is a classic example of creative civil disobedience, one of Kelly's categories of peace action described earlier. It is also a good further illustration of Palme's point about the need for people to put pressure on their governments. Whatever the future for Yesh Gvul, it is undoubtedly the only manifestation of the peace movement in Israel that has its establishment worried, to the extent that they have started to spy on and harrass Yesh Gvul leaders and increase the sentences for refuseniks. Undeterred, Yesh Gvul remains, in the best non-violent tradition, the unquiet voice of Israel's conscience.

Hildegard and Jean Goss-Mayr
(Austria/France)

Yesh Gvul is, of course, only one of very many examples around the world of people's non-violent refusal to submit to or co-operate with the unjust exercise of power. Two people who have dedicated their lives to the encouragement and enactment of such activities are Hildegard and Jean Goss-Mayr. Since the Second World War they have worked singly or together in every troubled region, and most war-torn countries in the world, and have become two of the most lucid and effective proponents of non-violence.

Their commitment to non-violence, like that of most of its well-known advocates such as Jesus, Gandhi or Martin Luther King, springs from a fervent (in their case Christian) faith. It was born or nurtured by experiences during the Second World War. Jean, born in 1912, was tortured by the Nazis as a prisoner-of-war, having been a decorated soldier in the French army; Hildegard, born in 1930, and her pacifist family, only narrowly survived Hitler's regime in Vienna.

Following the war they both became travelling activists for non-violence associated with the International Fellowship of Reconciliation of which Hildegard's father had been a co-founder, meeting and marrying in 1958. Since then they have worked extensively in the USSR and Eastern bloc countries, Latin America, Southern Africa, the Middle East, Asia, especially the Philippines, Spain and Portgual, Ireland, the UK, USA and Canada. Their work has met with some real successes. It helped to build in Latin America the movement for liberation and social justice through non-violence, Servicio Paz y Justicia, the leader of which in 1974 became Adolfo Perez Esquivel who later was to receive the Nobel Peace Prize. In the Philippines they were important in the nurturing of 'people power' which led in 1985 to the non-violent overthrow of the Marcos regime. The Goss-Mayrs' perception of the nature and role of non-violence is best conveyed through their own words, taken from Houver 1989:

Non-violence is based on the awareness of our ability to act out of love for justice, and in truth.

(p.35)

To say 'no' you must raise your head, you must show yourself, you must be a human being and not a slave. By saying 'non-violence' I begin by refusing a fatality: of violence, that is, of evil. . . . When I say 'no' to violence, I (Jean) act as a free man. I give man all his human dimension and in addition, I give him his divine dimension.

(p.1)

You just have to look around you to realise what methods all dictatorships throughout the world and throughout history use: lies, torture, political murder, hatred, injustice, exploitation of people and domination, open or latent. One word sums all that up: violence. Now what are the bases of all great revolutions? Truth, justice, solidarity, human rights, respect for men, love. It's easy to sum up: non-violence.

(p.116)

Non-violent action is fundamentally creative. It makes each particular individual discover the latent forces of truth within himself/herself and put them into action with imagination and initiative. Non-violence is a liberating force, quite the contrary to the forces and principles which govern our consumer society, which aim to control and dominate people's desires and actions. The true solutions to the conflicts and problems non-violent people have to face can only be found in communal action carried out by a basic group. In such a group, the creative contribution of each one is acknowledged and indispensable.

(p.105)

On the effectiveness of non-violence depends the survival of all humankind today. . . . We have reached a point of no return in the history of our species.

(p.73)

There will never be any disarmament if it stays on the level of diplomatic abstractions. . . . Disarmament is a struggle. In this struggle truth is by far the most important thing. We must tell the truth about armaments, about the effectiveness of arms in conflict, for protection and for profit.

(p.126)

If we don't want violence to overcome the whole planet, we must begin by rooting it out at home, in our own selves. The violence of each of us is our beam; we can see nothing as long as we accept it in front of our eyes. We can see nothing but 'violence' since everything will be perceived through it. Take the beam away and non-violence will become effective.

(p.76)

Every time we have recourse to violence, however slight it is, it is a step towards the abyss of total violence. Little wars only prepare the way for the big one, just as our little violences father the planet's violence. The non-violent movement is the property of no one, of no party and no group. The choice of non-violence demands the participation of everyone. The scientists in our modern industrial societies are responsible for our environment, the armed forces owe us the truth about armaments and their consequences, but they are also responsible for a non-violent civil defence. Non-violent commitment should be taken up by medical doctors and biologists. Whenever the major problems of our time are decided on, whenever there is an inevitable choice between degradation and absolute respect for human life, the presence of non-violent people is necessary.

(p.123)

Jean Goss died on 3 April 1991 as this book was in production.

Hans-Peter Dürr (Germany)

One scientist who acts as if he might have taken the Goss-Mayrs' injunction to heart is the German nuclear physicist Hans-Peter Dürr, Director of the Heisenberg Institute of Physics at the Max Planck Institute of Physics and Astrophysics, and Professor of Physics at Ludwig Maximilian University, both in Munich, Germany.

Dürr, born in 1932, is a quintessential interdisciplinarian. He has been professionally active in the field of energy policy (he has spoken and demonstrated against nuclear energy), science and responsibility, and epistemology and philosophy, as well as in his specialisms of elementary particle and nuclear physics. Recently he has also become very concerned about Third World, economic and ecological matters, and since 1985 has been a member of the Board of Greenpeace Germany. But his main campaigning work in the 1980s was on the theme of peace.

In 1983 he became a member of the Pugwash Conference and was co-founder of the Scientists' Initiative 'Reponsibility for Peace', leading to the Scientists' Peace Congress in Mainz, attended by 3,300 scientists, and the 'Mainzer Appell', a declaration against further nuclear armament. The following year another huge scientists' convention in Göttingen warned against the militarisation of space. Out of these conventions came interdisciplinary lecture series in 40 per cent of West German universities.

Dürr has been especially active on the issues of non-offensive defence and, his main preoccupation, SDI (the US 'Star Wars' Strategic Defense Initiative). A long article in *Der Spiegel* in 1985 argued overwhelmingly against the feasibility of the SDI concept. In 1986 he proposed a World Peace Initiative, of a similar scale to SDI, in which high technology would be reoriented to solve the problems of pollution, depletion, Third World poverty and economic injustice. This idea became established in 1987 as the Global Challenges Network, the methodology of which Dürr borrowed directly from SDI.

In presenting the vision of a complete defense against nuclear weapons, US President Ronald Reagan offered the American people a utopian dream which he called the Strategic Defense Initiative (SDI). But the proponents of that dream did not stop there. In the months after the President's announcement, the Fletcher Commission broke down that vision into hundreds of specific projects which seemed manageable at the individual, group or institutional level. While people working for peace do not believe in this dream, they can learn from the example.

A World Peace Initiative would undoubtedly generate a far broader and deeper consensus than SDI could ever hope to do. It could begin with the creation of a deliberative body somewhat like the Fletcher Commission. To it would be assigned the task of breaking the vision of peace into manageable portions. Taking 'peace' in its broader sense of environmental caring and resource sharing, as well as in the narrower meaning of reducing the weapons and institutions of war, the Commission could ask a myriad of smaller questions and identify a variety of tasks which could be undertaken. In many instances the Commission could probably identify those especially qualified to work on them.

(Dürr 1986, p.1)

The Global Challenges Network (GCN) meeting to identify specific projects took place in July 1987, when four projects were chosen: a measurement network; the Clean Baltic Sea Initiative; integration of 'nature' into price systems; and the preservation of local crops in their natural habitat. The first three of these projects were further developed at an international conference in November 1988. The Clean Baltic Sea Initiative has since brought together a number of environmental groups to tackle in a serious and integrated fashion the devastating pollution of this largely enclosed sea. The price systems project has become more widely focused on economic incentives to protect the ecosphere and is planning a major publication with international comparisons on this theme. The measurement network has established a computerised, decentralised system of monitoring the water quality of streams using bioindicators (living things which react sensitively to different environmental conditions).

According to Johannes Hengstenberg of GCN: 'The Clean Baltic Sea project demonstrates best the goals of GCN. It crosses barriers among disciplines, between environment and science and among countries and blocs' (Hengstenberg 1988, p.11). Such an approach is likely to become increasingly necessary if today's 'global challenges' are to be successfully addressed.

Johan Galtung (Norway)

Another scientist who has added a whole new dimension to the social sciences is Johan Galtung, often thought of as the founding father of peace research and the creator of several institutional vehicles for its pursuit. Galtung was born to Norwegian parents in 1930 and has had an international academic career spanning thirty years, five continents, a dozen major positions and over thirty Visiting Professorships, fifty books and more than 1,000 published monographs.

Turning to the social sciences after initial research as a mathematician, Galtung published his influential *Theory and Methods of Social Research* (Allen & Unwin) in 1967. In 1959 he had set up the International Peace Research Institute in Oslo, the first institute of its kind to make a mark in the academic world, and was its Director for ten years. In Oslo too he founded the *Journal of Peace Research* in 1964 and edited it until 1974. He was Professor of Conflict and Peace Research at the University of Oslo from 1969–77, during which period he also helped to found the Inter-University Centre in Dubrovnik, Yugoslavia, as a meeting place for East and West, and was for four years its first Director-General. High ranking university positions followed in succeeding years, interspersed with consultancies to the whole range of UN agencies: UNESCO, UNCTAD, WHO, ILO, FAO, UNU, UNEP, UNIDO, UNDP and UNITAR. Some of the subjects in which he held a Visiting Professorship in 1986 were international economics at Sichuan University, China, world politics of peace and war at Princeton University, USA, international studies at Duke University, USA, and Peace Studies at Chuo University, Japan. He is currently working on an integrated agenda with four main strands:

1. Comparative civilisation theory, exploring the underlying implications for peace and development of occidental and oriental civilisations;
2. The generation of textbooks in general peace theory and general conflict resolution;
3. Development theory, including issues of ecology, health and peace;
4. A new approach to economics which can more comfortably accommodate such major world goals as peace, development, human growth and ecological balance.

Galtung's analysis of the conditions for a peaceful world is based, appropriately, on an analysis of power, of which he conceives four distinct types: military, economic, cultural and, to co-ordinate these three, political power (Galtung 1987). Each type of power can be used for violent purposes, or it can be used to promote peace, which Galtung divides into negative and positive peace. This produces an eight-box matrix as in Figure 2, in which each box represents the possible contribution of a particular form of power to one of the two types of peace. Conversely, any particular conflict can be traced to an aggressive exercise of one or more types of power.

Example 1

The intervention of the USSR in Afghanistan can be seen as *military* aggression based on a desire for *cultural* domination through the imposition of communism.

Example 2

The intervention of the US in Nicaragua was first exemplified by *economic* aggression, through the US economic boycott of Nicaragua, followed by *military* aggression, through US support for the Contras, again deriving from *cultural* aggression based on the desire to impose capitalism.

	NEGATIVE PEACE	POSITIVE PEACE
MILITARY	*Military non-aggression* Non-*flow*, non-intervention Non-*stock* of offensive arms Defensive, non-provocative defence Transarmament Disarmament Abolition of war as an institution	*World peace-keeping forces* Non-violent intervention Stationing as buffers in crisis areas Stationing as hostages Co-operation in defensive defence World transarmament association World disarmament association World war abolition association
ECONOMIC	*Economic non-aggression* Nature, human, social, world *production* for basic needs *Distribution* to most needy SELF RELIANCE I Nationally: use local factors Locally: internalising externalities	*World economy* Nature, human, social, world *production* for basic needs *Distribution* to most needy SELF RELIANCE II Equitable exchange Symbiosis, mutual benefit Sharing externalities equally
POWER CULTURAL	*Cultural non-aggression* Participation in dialogue Not backed by military and economic power Criticise, internally, externally: • universalism and singularism • chosen people ideas • absolute cultural relativism	*World consciousness* World statistics, world images Conceptualisation and foreign policy as world domestic politics Positive views of humanity: • Multicentric space • Relaxed, oscillating time • More holistic, dialectic • Nature partnership • Equality, justice – inclusive • Minimum metaphysics
POLITICAL	*Internalise national interests* *Broaden democracy* National and local elections/party candidate *and* issue votes Nuclear-free municipalities Nuclear-free professions with Hippocratic peace oaths	*World institutions for world interests* *Broaden democracy* Chamber of governmental organisations Chamber of people's organisations Chamber of peoples World elections World referenda World service • environment • development

Figure 2 Visioning a peaceful world: how to weave states together, softening them, interlinking them
Source: Galtung 1987

In each case, solutions to such conflicts depend on acknowledgement by each superpower of the right of their neighbours to choose their own cultural (and economic) system and the consequent withdrawal of the military and economic power that is wielded against them.

It is to carry out the detailed application of such analysis to situations of conflict that Galtung has been actively engaged as a conflict resolution facilitator between North and South Korea, Israel and Palestine and East and West in Europe.

FROM NATIONAL TO GLOBAL SECURITY

At a time like the present, of important discontinuities and rapid change, projections about the future can be wildly inaccurate, but it is at precisely such times that such projections are most important if the threats and opportunities of such change are to be capably handled.

With hindsight it may be that the speech by President Gorbachev to the UN General Assembly on 12 December 1988 will be recognised as the moment when a wide public became aware that post-Second World War notions of national security had gone forever.

> The use or threat of force can no longer, and must no longer, be an instrument of foreign policy.... One-sided reliance on military power ultimately weakens other components of international security.... The concept of comprehensive international security is based on the principles of the United Nations Charter and is predicated on the binding nature of a new model of economic security, not through the build-up of arms ... but on the contrary through their reduction on the basis of compromise.
>
> (Gorbachev 1988)

Early the following year Michael Renner of the US Worldwatch Institute expressed the same reservations about a narrow emphasis on military security: 'The security of nations depends at least as much on economic well-being, social justice and ecological stability.... Pursuing military security at the cost of social, economic and environmental well-being is akin to dismantling a house to salvage materials to erect a fence around it' (Brown et al. 1989, p.133).

The UK's Saferworld Foundation broadened still further the concept of security:

The new definition of security must include international and transnational problems which have not previously been considered a part of security, but which threaten the well-being and the interests of states, societies and individuals as much as or more than military threats. These include: economic underdevelopment; overpopulation, environmental degradation; political oppression and violation of human rights; ethnic and religious rivalries; and terrorism and crime. Such non-military security concerns interact with the traditional military focus of security, causing or contributing to multiple levels of conflict and violence.

(Eavis 1990)

Thus expressed, the new security concerns are fully consistent with Galtung's (1985) identification of violence as the result of the denial of four basic areas of human need: for survival, welfare, freedom and identity.

One possible misunderstanding to be avoided is that a broader security focus will automatically provide easy answers to traditional questions of military security. As the Pugwash Conferences have declared:

Very little progress has yet been made in actually reducing the obscenely excessive nuclear arsenals. There is still the widespread belief that peace in Europe has been kept by the nuclear deterrent and that therefore nuclear weapons must be retained. Military research establishments are still busy inventing new generations of weapons while very little is being done to convert military industries to peaceful production. There is still the danger of military confrontation in several regions in the world, notably in the Middle East, with states acquiring the ballistic missile technology for delivery of weapons, and amassing chemical weapons as a substitute for atom bombs, as well as attempting to acquire nuclear weapons.

(Pugwash Council 1990)

The Pugwash Conferences themselves have been a valuable non-governmental contribution to peacekeeping since 1957. A small group of scientists, inspired by the Russell-Einstein Manifesto of 1955 which urged scientists to act in the face of the perils of nuclear war, met in the small Nova Scotia, Canada, village of Pugwash. By July 1990 173 such meetings had taken place and Pugwash was widely recognised as an influential and, at times of great Cold War tension, unique channel of communication, especially between East and West, on security issues. Its deliberations undoubtedly contributed to the Partial Test Ban Treaty (1963), the Non-Proliferation Treaty (1968), the Strategic Arms Limitation Agreements and Anti-Ballistic Missile Treaty of the 1970s, the Biological Warfare Convention (1972) and negotiations on chemical weapons and the Conference on Security and Cooperation in Europe (CSCE or Helsinki Process).

Despite its infancy, the main practical elements of the new security agenda can be clearly seen. Taken from Eavis and Clarke 1990 (pp.10ff) and elsewhere, these elements include:

- *Non-offensive defence*, based on the principle that greater security is derived from military deployments that do not threaten the security of potential adversaries. Such deployments of *defence*-focused forces seek to create a condition of 'mutual defence superiority' whereby both sides are capable of repulsing an attack but not of mounting one.

- *Minimum nuclear deterrence* (as a transitional objective) embodying an arsenal of only a few hundred nuclear warheads clearly designed only for retaliatory action coupled with a public commitment to no first use of nuclear weapons.

- *National nuclear disarmament* (as a final objective), by strengthening the Non-Proliferation Treaty's benefits for non-nuclear states, speeding the process of nuclear disarmament to minimum nuclear deterrence and beyond, putting in place a Comprehensive Test Ban Treaty to stop the development of the next generation of nuclear weapons; and greatly restricting the spread of nuclear reactor technology. A nuclear capability requires an expensive, sophisticated and specialised technological base and has already proved itself amenable to international detection. There is at least a chance that, under an enforced global prohibition on national nuclear weapons, backed up by an international nuclear capability, the very ability to make nuclear weapons would diminish. But even if this were not so, national nuclear arsenals should eventually be forbidden under international law. The fact that one cannot wholly deter a crime is no argument for legalising it.

- *Development of international institutions* for peacemaking, peacekeeping, guaranteeing security and building confidence. The most obvious current examples on which to build are the United Nations and the CSCE. With regard to the former, the experience of the Gulf Crisis showed that it was eminently possible to build a UN consensus against violations of international law. That experience needs to be generalised as soon as possible to ensure that all UN resolutions seeking to remedy such violations are acted on with comparable vigour. This would do much to deter violations in the first place.

The CSCE or Helsinki Process is a standing conference of the US, Canada and all European countries including the USSR except Albania. It was first convened in 1973 and led to the historic Helsinki Final Act in 1975, committing the signatories to respect for human rights and each others sovereignty, mutual co-operation and a follow-up process to monitor compliance. CSCE undoubtedly made a great contribution to the transformation in Eastern Europe. The next challenge is its development into a pan-European security institution, ultimately to replace NATO. The first ever meeting in Vienna in January and February 1990 of the military chiefs of staff of the CSCE nations indicates that such a development is not inconceivable. Moreover, there have been calls for a 'CSCE process' to be established elsewhere. The Italian government has called for a CSCM (Conference on Security and Cooperation in the Mediterranean region) and in November 1990 a group of two dozen recipients of the Right Livelihood Award suggested a CSCME (Conference on Security and Cooperation in the Middle East) on the CSCE model – continuing negotiations over a long period of time involving all interests and excluding no topic from discussion – to resolve that region's security problems. The CSCE example is clearly one of great potential.

- *Redeployment of expenditures* saved through disarmament to contribute to security in other areas. On realistic security assumptions, envisaging defence cuts of up to 50 per cent over ten years, such a 'peace dividend' from savings on NATO budgets alone over the 1990s could amount to US$1.4 trillion (Chalmers 1990, p.98). Such a sum could be invested in increasing non-military security in three different ways: conversion of defence industries to civilian production; modernisation and 'greening' of domestic economies; debt cancellation, appropriate technology transfers and environmental regeneration to benefit the Third World countries. These investments will need careful planning and monitoring. There is now sufficient experience with conversion, for example, to know that it needs to be enacted within a coherent national strategy and that, initially, it is not a cheap option:

> If one may expect substantial financial benefits, the so-called peace dividend, in the long run, there is an intermediate period of several years when there is in fact a peace penalty. Money has to be injected into the industries to construct new buildings, make new equipment dedicated to civilian production. This is also the case when equipment or personnel are transferred to the civilian sector.
>
> (Journe 1990, p.2)

Nor is conversion purely a technical matter. Defence industries need careful preparation for the civilian market. As Michael Renner writes: 'Conversion goes beyond a mere reshuffling of people and money. It involves a political and institutional transformation' (Brown *et al.* 1990, p.157). However, the political benefits of conversion are immense in terms of employment, because civilian sector spending employs up to seven times as many people as defence; and in terms of the opportunities offered for industrial restructuring towards conservation and efficiency.

Internationally, too, the peace dividend offers the best hope for igniting a true development process. Debt cancellation is a top priority:

> In one way or another the developing world's debt will need to be effectively written down, case by case and in an orderly manner, by up to 50 per cent over the next five years, including an outright cancellation, or the equivalent, of the remaining debts of the very poorest and most debt-burdened developing countries.
>
> (UNICEF 1990, p.25)

An effective way of doing this would be the use of the peace dividend to purchase discounted government debt on secondary markets as described in UNICEF 1990 (p.64). But here, as with other transfers to the Third World, care will need to be taken that debt cancellation does not provide the opportunity for another spending spree for the Third World elites. Susan George (1988) has recommended that the debt cancellation be tied, in a '3-D solution', to environmentally sustainable development and progress towards democracy. Such conditionality is vital if the peace dividend's historic potential to fund the transition from national to global security is to be realised.

CONCLUSION

Olof Palme in the Introduction to his Commission's report made a moving statement of the vision that underlay their work.

> Our vision is of an international order where there is no need for nuclear weapons, where peace and security could be maintained at much lower levels of conventional armaments and where our common resources could be devoted to providing greater freedom and a better life for people.
>
> (ICDSI 1982, p.xiii)

It is clear that while governments may, and need to, sign treaties which express this vision, they will not do so until it has first become widely established in their populations and been articulated and campaigned for by people's organisations in their countries along the lines described here. Vaclav Havel in 1984 expressed the same point thus:

> The perspective of a better future depends on something like an international community of citizens which, ignoring the state boundaries, political systems and power blocs, standing outside the high game of traditional politics, aspiring to no titles and appointments, will seek to make a real political force out of a phenomenon so ridiculed by the technicians of power – the phenomenon of human conscience.
>
> (Havel 1988)

Realising Palme's vision, giving political power to the human conscience: such is the historic mission of the world's peace movement at the end of the second millennium.

4

IN DEFENCE OF
HUMAN RIGHTS

Sometimes what is *not* done can be as significant as what is. As has
been discussed the UN system has established high-level commissions
to investigate three of the four great 'holocausts' currently afflicting
humanity – war, poverty, environmental destruction – and their
reports have done much to stimulate international concern and
debate on these issues and even some action. But there has been
no such high-profile special commission on the fourth inter-human
holocaust: the systematic infringement of the most basic human
rights of their citizens by most of the governments of the world.
As Amnesty International reported in 1989:

> Tens of thousands of people were deliberately killed in 1988
> by government agents acting beyond the limits of the law. . . .
> Killing grounds were many and varied. . . . Some (people) were
> killed in full public view, others in secret cells and remote
> camps. Some victims were shot down near battlefields, others
> in mosques and churches, hospital beds, public squares and busy
> city streets. Prison cells and courtyards, police stations, military
> barracks and government offices were all sites of political killing
> by agents of the State. Many people were killed in their own
> home, some in front of their families.
>   Victims were assassinated by snipers, blown up by explosive
> devices or gunned down in groups by assailants using automatic
> weapons. Others were stabbed, strangled, drowned, hacked to
> death or poisoned. Many were tortured to death. In Colombia,
> Guatemala, El Salvador, Syria and the Philippines victims were
> often severely mutilated before they were killed. Their bodies
> were burned or slashed, ears and noses were severed and limbs
> amputated.

Many governments used methods of torture that were inherently life-threatening, such as beating, electric shocks, drugs, immersion and hanging. In 1988 deaths after torture were reported in Turkey, El Salvador, Indonesia, Iraq, China, Syria and Burma.

(Amnesty International 1989, pp.9–10)

The Universal Declaration of Human Rights in the founding Charter of the United Nations, to which all member-states are signatories, forbids all such activities, of course. The Universal Declaration can also probably be said to be the bedrock on which the UN itself is founded. Certainly the UN would be a quite different institution without it. So how is it that such gross violations of this organisation's single most important utterance occur routinely year by year without investigation and retribution by other member states who never tire of saying how much they deplore such abominations?

The first part of the answer to that question is that, despite the UN Charter's initial preamble of 'We the people of the United Nations . . .' the UN is actually an organisation of governments and not of 'peoples'. The second part of the answer is that many governments are not only not representative of their people but actually seem prepared to wage war against them if such seems to be in the interests of the ruling elite. And the third part of the answer is that there are not enough governments, even among those that are broadly democratic, that are prepared to prosecute with vigour, at the UN and elsewhere, the cause of human rights and their abuse, for the violating governments to feel the heat of world opinion.

Worse, some of the governments which should have been in the best position to promote the observance of human rights abroad have deeply compromised themselves by seeking strategic or economic benefits on the back of human rights violations. Thus the absolutely justified US condemnation of Soviet human rights abuse domestically and abroad came across to the international community as little more than ideological point-scoring, because the US was simultaneously backing some of the most bloody regimes in Latin America, including Guatemala and El Salvador, both specifically indicted in the quotations from the Amnesty report above. Similarly the UK permitted its economic interests to fly in the face of calls by practically the whole world's black community for economic sanctions against South Africa, and sold weapons not only to Chile's Pinochet regime but also to Argentina's

Galtieri government at the height of the Argentinian disappearances and shortly before the Falklands War (CAAT 1981).

It seems extraordinary that the United Nations has not yet spawned a group of governments committed to campaigning within the UN for the observance of the Universal Declaration of Human Rights, rather as the Western countries sought successfully to use the Helsinki Final Act to put pressure on the USSR and its allies. As will become apparent, such campaigning groups of citizens exist within practically all countries, often at enormous hardship to their members, to press for their government's recognition of human rights. Is it too much to hope for that one of the fruits of the new climate in East–West relations may be a new commitment by the superpowers, their allies and democratic governments everywhere not only to human rights observation at home, but their vigorous promotion abroad, in support of these courageous citizens who actually take the UN Declaration of Human Rights seriously?

Theo van Boven (Netherlands)

Despite its apparent inactivity on human rights issues, the United Nations *does* actually have a permanent Commission on Human Rights, so that it possesses the institutional means to give the necessary weight to these matters. Unfortunately, institutional means without governmental commitment do not allow much to be done, as former Director of the UN Division on Human Rights, Theo van Boven, found to his cost.

Van Boven was born in the Netherlands in 1934 and obtained a doctorate in law in 1967. He was for ten years until 1977 a lecturer in human rights at the University of Amsterdam and from 1970–5 was the Netherlands' representative on the United Nations Commission on Human Rights. From 1977–82 he was Director of the UN Division of Human Rights, since when he has been Professor of Law at the University of Limburg. In addition, van Boven has served on numerous councils and committees dealing with human rights, including the Council of the International Institute of Human Rights (France) and the European Human Rights Foundation (UK), of which he is the Chairman.

As Director of the UN Division of Human Rights van Boven argued consistently that concern for human rights should not be a marginal activity within the UN system, but should become the core element of development strategies at all levels. He sought to break through the selective approach of the UN in human rights matters, and to deal more consistently with the gross violations of human rights in a large number of countries on all continents, including enforced and involuntary disappearances, torture, summary and arbitrary execution, and discrimination against indigenous peoples. He contributed to the creation of fact-finding mechanisms in these areas in order to bring pressure on defaulting authorities and to provide relief to victims. He was concerned also to identify the root causes of human rights violations in connection with the development process, patterns of economic and political domination, militarisation of societies and racial discrimination. In addition, he worked hard to strengthen the links of his office with non-governmental organisations.

His uncompromising approach to these matters led to major policy differences with the UN Secretary-General which led to his UN contract being terminated in May 1982. An indication of what the world community lost with his departure was provided by a paper he wrote in 1989 'The international human rights agenda: a challenge to the United Nations' (van Boven 1989), which set out both the current role of the UN on human rights issues and an agenda to make it more effective. The UN's current role van Boven classifies under three headings:

- Standard setting, through the vitally important International Bill of Rights and other human rights pronouncements.
- Protection, through finding ways of making governments accountable to the human rights standards set.
- Promotion, through education and training on the importance of the rule of law and respect for human rights.

Van Boven's five-point agenda for the future focuses first on the need to improve the accountability of governments for human rights abuse. Van Boven sees the need for the involvement of both independent experts and non-governmental organisations (NGOs) in this process for it to be effective. Two other points of his mesh closely with arguments to be made later in this book: that there need to be far tighter procedures for the protection of the rights of indigenous people; and that human rights imperatives need to be closely woven into the development programmes of UN agencies and others. As will become clear, too often 'development' results in the pauperisation or complete ruination of large groups of people. Finally van Boven asks for two institutional reforms: more coherence in standard setting to cover the requisite areas without ambiguity or duplication; and the provision of more resources to the UN human rights programme, which currently accounts for just 0.7 per cent of the UN budget, and which is hopelessly overstretched in terms of the demands made upon it.

Assuming this eminently reasonable agenda is similar to that promoted by van Boven within the UN, it is a sad reflection of human rights commitment among UN members and government officials that it made the man promoting it *persona non grata* within the organisation.

Amnesty International

The UN Commission on Human Rights could publish every year a hard-hitting report on human rights abuse around the world. It could pressure governments directly to improve their record, and could organise networks of people (in the spirit of 'We, the people') round the world to do the same. These same people could communicate directly with wrongly imprisoned or threatened citizens worldwide to give them direct support and assure their governments that the UN, and the world, was watching. The UN Commission could do all these things, but does not. Amnesty International (AI) does.

AI was founded in 1961 following a letter from a lawyer in a British Sunday newspaper calling for an international campaign to protect human rights. The letter drew 1,000 responses. Within twelve months the new AI organisation was active in seven countries and had taken up 210 cases. In 1989 AI had more than 700,000 members in over 150 countries and more than 4,000 groups worldwide. It had investigated more than 32,000 cases of individual prisoners of conscience. In 1988 1,566 prisoners under AI adoption or investigation were released (though AI is careful not to claim direct credit for this). In 1990 AI worked on more than 3,200 such cases, involving more than 4,500 individuals from countries worldwide (personal communication, AI office, London, 16 April 1991).

AI's research is done by a 250-strong International Secretariat based in London with stringent rules for cross-checking allegations of and information about human rights abuses. On the basis of this information its members write and give support to prisoners and lobby offending governments (in urgent cases through the Urgent Action Network which sends telexes or telegrams to government decision-makers). At international level, AI works for greater commitment to human rights at the UN and in other inter-state fora. AI national sections do the same within their own country.

AI's essential position on human rights issues is very simple. It aims

- to seek the release of prisoners of conscience. These are men, women and children who are detained because of their beliefs, colour, sex, ethnic origin, language or religion who have not used or advocated the use of violence.
- to work for fair and prompt trials of all political prisoners and on behalf of political prisoners detained without charge or trial.
- to oppose the death penalty and torture or other cruel, inhuman or degrading treatment or punishment of all prisoners without reservation.

(Amnesty International undated, p. 1)

In 1989 AI's central budget for this work was less than £10 million, perhaps a third of the organisation's total outlay. This figure reveals two things: the extraordinary cost-effectiveness of Amnesty's operation; and the extraordinary scale of priorities which allows the world's foremost human rights institution to subsist on an annual budget equivalent to some ten minutes-worth of total annual expenditure on weapons.

International Rehabilitation and Research Centre for Torture Victims (Denmark)

The condemnation and abolition of torture has been a central concern of Amnesty International since its inception and is the target of an ongoing AI campaign which started in 1984. Eleven years earlier AI had appealed to the medical profession to help fight torture and in response Dr Inge Kemp Genefke formed the first Amnesty International medical group in Denmark. Its pioneering investigations into torture and its consequences for its victims led to the establishment of more medical groups, and by 1982 there were twenty-nine such groups with over 4,000 doctor-members. The need for treatment and rehabilitation of torture victims then led in 1982 to the establishment in Copenhagen of the International Rehabilitation and Research Centre for Torture Victims (RCT), with Dr Genefke as Medical Director.

RCT has the following objectives:

- to operate a centre for rehabilitation of persons who have been tortured, and of such persons' families;
- to instruct Danish and foreign health-service personnel in examination and treatment of persons who have been tortured, and through instruction in wider fora to propagate knowledge of torture, forms of torture and the possibilities of rehabilitating persons who have been tortured;
- to conduct and initiate research into torture and the nature and extent of consequences of torture;
- to operate and extend an international documentation centre;
- and through the above activities to contribute to the prevention of torture.

RCT now has a staff of thirty-five consisting of doctors, psychotherapists, physiotherapists, nurses, social workers, interpreters, librarians and technical administration personnel. It also has permanent arrangements with a number of other specialised physicians, psychotherapists, dentists, interpreters and translators. In addition RCT has organised international seminars in Denmark, the Philippines, Kenya, England and Uruguay, and has helped establish centres in many countries round the world.

RCT's research programme has contributed enormously to understanding about the treatment of torture victims and has received much publicity worldwide. To aid the process of disseminating this information the RCT established in 1987 its International Documentation Centre, to include papers, reports, press cuttings and audio-visual material about every aspect of torture and its consequences, much of which is not registered anywhere else.

An important focus of RCT is doctors, ethics and torture. Doctors have been estimated to participate in more than 60 per cent of torture-cases. The experiences in Uruguay, Argentina and Chile all confirm how difficult it is to bring these doctor-torturers to justice even after democratic governments have been restored (RCT 1990). The first international meeting on this subject was held in Copenhagen in 1986, organised jointly by the Danish Medical Association and the RCT. A follow-up meeting on 'Doctors involved with torture' was held in Uruguay in 1987, organised jointly by the Uruguayan Medical Association and the RCT. An important result of these meetings has been the Statement of Madrid of November 1989, in which the medical associations of the EC agree to fight against doctors' participation in torture and integrate the subject of torture in the medical educational curricula.

An international symposium on 'Torture and the Medical Profession' was held in Tromsø, Norway, in June 1990, where important recommendations were adopted on registration and monitoring of established cases of medical involvement in torture; establishment of national and international tribunals for such cases; and adoption of the Statement of Madrid by all national medical associations.

RCT also hosted in 1986 and 1988, in collaboration with the Dutch government, meetings with the World Health Organization, the outcome of which was that WHO resolved to seek the introduction into medical curricula of the nature and effects of torture, as is already the case in the University of Copenhagen.

In pursuing its overall aim of contributing to the prevention of torture, RCT is in close contact with both the UN Committee Against Torture and the European Committee for the Prevention of Torture; its former chairman Professor Bent Sørensen is a member of both bodies.

HURIDOCS

In one of its factsheets Amnesty International notes that there are now over 1,000 domestic and regional organisations working round the world to protect human rights; the Master List of the Human Rights Internet, an international communications network and clearing-house on human rights, based at the University of Ottawa, gives names and addresses for over 2,000 such organisations. As AI itself has shown, the single most crucial factor for the successful operation of all these organisations is information, both in terms of accuracy and timeliness. It is to the improvement of performance in this area that one of the most interesting new human rights initiatives of recent years is dedicated, the Human Rights Information and Documentation Systems, International, HURIDOCS.

HURIDOCS was founded in 1982 and is now a global network with a participation of several hundred human rights organisations.

It was founded on the basis of a recognition that:

1. Information was not available to those who could best use it;
2. With the development of modern technology information was becoming less and less accessible to the defenders of human rights and more and more a weapon in the hands of the violators of human rights.
3. There was a lack of systematic treatment of information and documentation in human rights organisations;
4. There was a need to build a communications network.

(HURIDOCS 1989)

HURIDOCS' aim is to improve access to and dissemination of public information on human rights, broadly defined, and including civil and political as well as economic, social and cultural rights. This it seeks to do by imparting more effective, appropriate and compatible methods and techniques of information handling, working in particular to strengthen South-based human rights groups' competence in this regard.

HURIDOCS does not itself collect documents but seeks to establish a decentralised network of organisations concerned with documentation and information. Believing that networking is the only way to deal with the growing flow of human rights information, it tries to provide the necessary infrastructure which will lead to growing competence and professionalisation within organisations engaged in its network. HURIDOCS seeks to involve in its network organisations with manual as well as with computerised documentation systems.

HURIDOCS offers its participants basic tools for information handling and documentation control; development of new standards for information recording; teaching and training in information handling techniques; advice on the establishment and strengthening of information systems; limited advice on computer software and hardware; and co-ordination among the different documentation centres. HURIDOCS' training is in-depth and thorough. Its course on 'Human Rights Information Handling in Developing Countries' in the Philippines in December 1988 lasted six weeks and was an impressive testimony to the new professionalism among human rights organisations.

Since then HURIDOCS has organised a large number of shorter courses and workshops and is continuing such work in the future. An example was the Latin American HURIDOCS workshop

'*Segundo Mortes*' held in May 1990 and named after the Salvadorean Jesuit priest who had been an active member of the HURIDOCS International Advisory Council and was murdered in San Salvador with five Jesuit colleagues and two other collaborators in November 1989. The workshop created a Regional Working Network of Latin American human rights organisations.

Universal human rights are, as the name implies, applicable to all people, but there is one part of humanity for whom such rights are overwhelmingly honoured in the breach rather than the observance: women, comprising half the world's population; and the world's approximately 200 million indigenous, sometimes also called native or tribal, peoples.

WOMEN

Women are half of humanity. They do two-thirds of the world's work. But they earn perhaps one-tenth of the world's income; and they own less than a hundredth of its property (UN 1979). Most of the work they do is not paid at all and is not even counted as part of the world's wealth, despite the fact that this unpaid work: bearing, nurturing and educating children; homemaking; caring for the sick and elderly; farming for subsistence, fetching and carrying wood and water – this work is the very foundation of human wealth-creation. Yet economists, including those in such places that are supposed to be promoting women's rights, such as the UK Equal Opportunities Commission, call the women who do such work 'economically inactive' (EOC 1989). When women do take paid jobs, they are likely to be in areas of relatively low pay and low security; and even when they do the same work as men, they are likely to be paid less for it. And, of course they will still, overwhelmingly, also be responsible for the housework when they get home at the end of the day.

If the terms of women's productive work are bad, those of their reproductive work are, if anything, worse. They can be subject to barbaric practices like female circumcision. In forced marriages they are considered the personal property of the family head, to be given away as he sees fit; or in some societies, sold or, in others, only disposable of on payment of a dowry. Once in marriage they can be, and very often are, mercilessly exploited, beaten or raped with impunity. This can happen outside marriage too; and of course there is the further possibility of buying a women's body without her person. In South East Asia the sex industry is a big foreign exchange earner. One million Japanese and other businessmen in 1988 were estimated to go to Thailand, the Philippines, South Korea, Taiwan and Hong Kong on sex tours. In Bangkok alone there are between 100,000 and 200,000 prostitutes, many of them girls of 15 or 16 or less, sold by their destitute parents to bars that keep them chained to their beds when they are not working (Waring 1989, p.152).

It might have been imagined that economic development would have ameliorated these conditions. Unfortunately, there is significant evidence that, in fact, it has contributed to making women's situation worse. One of the 'fundamental findings' of Janet Momsen's study *Women and Development in the Third World* (Momsen 1991) was that 'economic development has been shown to have a differential impact on men and women and the impact on women has, with few exceptions, generally been negative' (p.4). Momsen identified some of the key issues as follows:

> Modernisation of agriculture has altered the division of labour between the sexes, increasing women's dependent status as well as their workload. Women often lose control over resources such as land and are generally excluded from access to new technology. Male mobility is higher than female, both between places and between jobs, and more women are being left alone to support children. Women in the Third World now carry a double or even triple burden of work as they cope with housework, childcare and subsistence food production, in addition to an expanding involvement in paid employment. Everywhere women work longer hours than men.
>
> (p.1)

Because of the importance of women's role in Third World agriculture, Momsen analyses its modernisation in particular detail through the various changes in the rural economy that are involved and their impact on women's socio-economic condition with regard to property ownership, employment, decision-making, status, level of living and nutrition and education. None of the changes are unequivocally positive; several are unequivocally negative, with the exception of possibly increased educational opportunities. As with agriculture, so with industry, according to Madhu Bhushan (1989) in India.

> Technological advancement has proved to be no boon to Indian women. . . . Their position in the industrial sector has

deteriorated drastically. As wages and working conditions have improved, men have taken over the jobs traditionally held by women in the textile and jute industries, and in mining. . . . All these trends that 'modernisation' and 'progress' have brought lead us to conclude that women increasingly are being victims of new forms of direct and structural violence.

(p.27)

This systematic erosion of women's role in economic life leads Momsen to speculate about a pervasive marginalisation of female employment in the development process, with four dimensions:

Firstly women are prevented from entering certain types of employment, usually on the grounds of physical weakness, moral danger, or lack of facilities for women workers. Secondly, marginalisation can be seen as 'concentration on the periphery of the labour market' where women's employment is predominantly in the informal sector and the lowest-paid, most insecure jobs. Thirdly, workers in particular jobs may become so overwhelmingly female that the jobs themselves become feminised and so of low status. A fourth dimension is marginalisation as 'economic inequality'. This aspect refers to the economic distinctions which accompany occupational differentiation, such as low wages, poor working conditions and lack of both fringe benefits and job security in work thought of as 'women's'.

(ibid. p.68)

'Marginalisation' is a term that is often used for Third World countries' experience of the global economy. Birgit Brock-Utne (1989) makes an explicit comparison between the relationship of women *vis-à-vis* men, and that of the Third World *vis-à-vis* industrial countries, elaborating on four aspects common to both: the paternalism in the relationships; the desire of radical feminists and Third World radicals to change the dominant system and not simply win more equal terms within it; the prevalence in both situations of single-track mechanistic solutions to the problems involved; and the persistence of relative dominance of men over women and the First World over the Third World even where absolute improvements in women's and Third World conditions have been made.

This book cannot do justice to these immensely complex, multi-levelled issues of economic and other discrimination against women, not least because such justice can only be, and has often been, done by women. The three references that follow are in no sense

75

'representative' of the vast literature on and by women, but they open doors onto different parts of it, and onto other areas, which can be followed up if desired. On the political economy of women's production and reproduction there is Marilyn Waring's *If Women Counted; A New Feminist Economics* (1989). For a positive and forward looking assessment of women and development worldwide, there is Hilkka Pietilä and Jeanne Vickers' *Making Women Matter: the Role of the United Nations* (1990). And, at a personal level, there is Sylvia Ann Hewlett's academically precise and heart-wrenching account of a successful career-woman-cum-mother making it, and not making it, in the man's world of the US, *A Lesser Life* (1986).

While this book cannot present a comprehensive analysis of female oppression, it would certainly not be complete without at least a partial description of women's response to and struggle against it. Berenice Carroll (1989) has presented a historical survey and classification of women's direct action which traces women's protests back to ancient times. Their methods exhibit immense variety in all kinds of actions – constitutional, constructive, symbolic, economic and industrial – including physical intervention, non-cooperation and civil disobedience. This variety is, of course, still very much in evidence today, and this section will end with two case-studies of women's groups working in very different ways to benefit large numbers of women subjected to very different oppressions. The second is a broadly-based human rights and development group from Costa Rica, CEFEMINA. The first is challenging one of the most oppressive expressions of male power: the patriarchy espoused by much of fundamentalist Islam.

Women Living Under Muslim Laws (WLUML)

One-fifth of all women, about 500 million people, live under the jurisdiction of 'Muslim' laws in some three dozen countries of the world. One such country is the Iraq of Saddam Hussein, whose Revolutionary Command Council in 1990 decreed that 'any Iraqi who kills even with pre-meditation his own mother, daughter, sister, niece or cousin for adultery will not be brought to justice' (WLUML 1990). In Pakistan the Hudood Ordinance promulgated in 1979 has resulted in raped women being whipped and imprisoned (WLUML 1987). In Iran between 1979 and 1985 (and more since) over one hundred women were stoned to death for alleged adultery, and thousands more jailed, tortured or executed (WLUML 1989).

WLUML's first task and initial achievement has been to show women in these different countries that, in fact, these 'Muslim' laws differ from and sometimes contradict each other; that they often derive from highly questionable readings and interpretations of the Koran; and that they have more to do with the exercise of political power and patriarchy than theology. One of the WLUML documents states: 'Depriving us of even dreaming of a different reality is one of the most debilitating forms of oppression we suffer' (WLUML 1988). Changing the perception of ineluctable religious 'truth' is the first step toward regenerating such dreams.

It is achieved by a combination of rigorous and professional research and effective communication. WLUML's network of researchers is analysing family laws, personal codes, marriage contracts and other legal instruments in twenty countries, to be published in 1992 as a handbook on law for women in the Mulsim world. At the same time another group is formulating an Islamic 'liberation theology' which, as in the Christian case, will stress the liberatory strands in the great religion.

These are longer-term projects. On a regular basis WLUML publishes dossiers of discussion and case-studies on women's legal issues from different countries. They issue Alert for Action briefings, from which the oppressive facts above were taken, on issues of special urgency. And they organise personal exchanges between Muslim women from different countries to enable them to share experiences directly.

WLUML has only been going since 1984/5. Its Algerian founder Marie-Aimée Hélie-Lucas co-ordinates the international secretariat from France, with other co-ordinators in Pakistan and Sri Lanka. This core group co-ordinates a larger group of women each working autonomously alone or in sub-groups on research, training, campaigns or information exchange. These two groups in turn are in touch with large numbers of individual women and women's groups from twenty-seven Muslim and many other countries. Their co-ordinating budget in 1990 was US$160,000.

WLUML has already had some successes: a woman kidnapped in France for a forced marriage in Algeria was returned and the marriage annulled after a WLUML campaign; eleven other North African women have been liberated from forced marriages; on the basis of WLUML briefings, a women's group in the UK persuaded the Home Office not to deport a woman to Pakistan to face an utterly unjust charge under the Hudood Ordinance; WLUML has blocked the deportation of three other women and obtained the return to their mothers of several children who had been kidnapped by their fathers. But the real flowering of this initiative is still to come, as one of the most oppressed groups of people in the world comes to realise the arbitrary nature of their oppression; experiences the solidarity of others determined to resist it; and perceives that a different reality can not only be dreamed. It can be achieved.

CEFEMINA (Costa Rica)

CEFEMINA, meaning the Feminist Centre for Information and Action, was founded in 1981 but has its roots in the Movimento Para Liberacion de la Mujer, which had been active since 1974. CEFEMINA describes itself as 'a private non-profit organisation dedicated to public service. Our work is oriented towards improving the overall well-being of women, beginning with the problems of daily life' (CEFEMINA undated).

CEFEMINA is active in many different areas, including:

- Housing: CEFEMINA has helped women to participate in the planning of their communities and build over 3,000 houses in different parts of the country. The most successful of these projects, at Guarari is described in more detail below.
- Healthy living: here 'the communities organise themselves, with the help of health professionals, to assist and train members of the community in different aspects that pertain to health: reproduction rights and family planning, prenatal control, breast-feeding, child development and early childhood stimulation, nutrition and physical well-being, recreation and sports, sexuality, medicine and drug-abuse and the dangers of contaminants and chemicals, defense of the consumer and warnings about the unscrupulous activities of certain industries, such as baby food and others, the resurgence of the use of traditional medicine and information on AIDS and other diseases' (CEFEMINA undated). In 1987 CEFEMINA organised the Fifth International Meeting on Women and Health, which drew 1,000 participants from eighty-two countries.
- Women's protection, through a programme called 'Women, you are not alone', which is a permanent network operative throughout the country to help abused women and prevent violence in

the family or community. CEFEMINA is also the national headquarters for the Latin American and Caribbean Network for the Prevention of Violence against Women.

- Women and justice, including legal advice and education and training on women's legal rights.
- Women and environment, including many grassroots, community-based environmental actions, the organisation of the 'First Congress on Women and Forest Action Plans in Central America', and the co-ordination of the 'Women and Sustainable Development in Central America' programme (see below).
- Publication of a magazine *Mujer*, a forum of dialogue and debate on sensitive subjects for women, and other publications. CEFEMINA also has a library relating to women's health.

Probably CEFEMINA's most important achievement to date has been its participation in the Guarari community self-build low-cost housing project. Following struggles against bad public housing practices, and earlier self-build experiences, CEFEMINA and a self-build housing group, COPAN, were allotted 118 hectares of land to develop into an urban community. 3,000 houses are planned, carefully integrated into the environment and preserving, indeed enhancing, a water-course running through the site. With expert advice, the houses are being planned, designed and built by the low-income communities that will live in them. To qualify for a house, a person must do 900 hours voluntary work on the project. The house is then paid for over fifteen or twenty years.

The women leading the process in the communities are also involved in CEFEMINA's other programmes, including health care and income generation. There is a special emphasis on keeping the environment green, with carefully tended gardens, tree planting and herb growing. The project is being watched carefully by the Costa Rican government as a possible model for replication. So far it has been extremely successful, with 1,000 houses built, costing up to 30 per cent less than similar publicly funded houses elsewhere and resulting in an immeasurably improved environment. An IUCN evaluation concluded: 'Experience at Guarari is showing that this type of community planning is cost effective and brings additional benefits to both the inhabitants and the land they occupy' (IUCN 1989). The Executive Director of the Arias Foundation for Peace and Human Progress has written: 'Guarari is now a model community in Costa Rica. Many representatives of national and foreign organisations have visited and used it as the basis for new projects throughout the region' (personal communication, 16 May 1991).

CEFEMINA is also co-ordinating a major project with IUCN on 'Women and Sustainable Development in Central America'. On the premise that 'Woman is the principal figure in the managing of natural resources, especially in developing countries' and 'The women's groups have shown themselves to be informed and are energetic and effective agents for conservation', this project is seeking to link a number of women's groups in all countries of Central America. The objectives of this ambitious programme are as follows:

To discuss and generate a theoretical framework for the Central American program on Women and Sustainable Development, with a prioritization of the areas of interest regarding women and the management of natural resources and that takes as an objective the improvement of the quality of life for women and their families.

To promote mechanisms of communication and exchange of information between the diverse groups of women with specific interests in Central America that can support and promote the participation in the rational management of natural resources.

To prepare educative and informative materials directed at women that can be used for training and education programs.

To train various Central American women's groups in the priority areas of interest in the management of natural resources and appropriate technologies.

To compile information regarding traditional knowledge of women, the management and use and exploitation of natural resources.

(Solis & Trejos 1990)

CEFEMINA and IUCN contacted a number of women's groups in Central American countries, focusing on a wide range of issues including health and environment, and were put in touch with a number of grassroots groups, including many craft groups which use natural resources. It is mainly with these groups that the programme will work.

INDIGENOUS PEOPLES

Since the beginning of European expansionism, followed by Western imperialism and continued with the 'development' policies pursued by practically every government in the world, tribal peoples have been dispossessed, enslaved and exterminated. The story has been the same for the North American Indians, displaced from a whole continent by white incomers; South American Indians, mercilessly assaulted by the Spanish temporal and religious power; Australian Aborigines hunted down by the British; and many other population groups worldwide. That these peoples have survived at all in the face of such rapacious hostility is a remarkable testament to their courage and the tenacity of their culture.

Survival International

Since 1969 indigenous peoples have not been totally without organised support in the industrial world for that was the year that saw the foundation of Survival International.

Survival International is a worldwide movement to support tribal peoples. It stands for their right to decide their own future and helps them protect their lands, environment and way of life. The initial focus for its work came from genocidal attacks on Amazonian Indians in Brazil. Initially it concentrated on researching and supporting concrete projects with tribal peoples, especially in the health and education field, and this is still an important part of its work. But its emphasis has shifted over the years to campaigning and public education, modelled in some respects on Amnesty International.

Thus Survival's 10,000 members are spread throughout the world and organised into national sections where their concentration or enthusiasm so warrant. Survival also makes effective use of the Urgent Action mailing technique pioneered by Amnesty, seeking to put hundreds or thousands of letters onto the desks of decision-makers responsible for the threat to tribal people. Survival also has the same embargo on government support for its work, in order to preserve its independence, which Amnesty, with very few clearly defined exceptions, also maintains. And Survival is also very careful to share responsibility for successes with the indigenous organisations with whom it campaigns.

Successes there have been. In the past two years:

- NATO decided against building a second air base on Innu land in Labrador, Canada, following continuous protests by the Innu and Survival's letter campaign.
- Survival's campaign against a new pulp mill on tribal land in Indonesia by Scott Paper, the largest tissue paper business in the world, led to the company's withdrawal. This saved the forest homes of 15,000 tribal people from destruction.
- By lobbying the Botswana government, Survival helped prevent the expulsion of the Bushmen and Bagkalagadi people from the Central Kalahari Game Reserve.
- The auction of a preserved head of a Maori warrior by UK auctioneers was halted and this brought the cancellation of further sales of human remains by other auction houses.
- The Indian government cancelled a huge hydroelectric dam on the Bodhghat river, saving the lands of 10,000 tribal people.

These results came from work on fifty specific cases in about twenty-five countries. A major focus of 1990s campaigning was the Yanomami Indians in Brazil, whose lands have been invaded by over 45,000 gold prospectors (*garimpeiros*). Over the past two years, an estimated 15 per cent of the Yanomami population have died, largely from diseases brought in by the goldminers. Survival hosted the first visit ever from a Yanomami representative to Europe and this helped to bring the Yanomami situation to the attention of the world. Since Davi Yanomami's visit in December 1989, Survival supporters have raised over £70,000 for emergency medical aid. In some communities where the health care has been sustained, the rate of malaria is down from 90 per cent to 10 per cent of the population. All this work was carried out within an annual budget for the organisation of less than £500,000.

Perhaps even more impressive and important for the long term than the individual successes of campaigns for tribal peoples' rights has been the unmistakable shift in public perceptions of tribal people. It is still unfortunately possible to find blatantly racist assertions such as that from the Gujarat Chief Minister who said of the culture of the tribal people due to be destroyed by the Narmada dams:

> I do believe that tribal culture should be preserved, but it should be in the museum, not their real life.
>
> (quoted in Survival International 1989, p.4)

or from the Sarawak Minister of Tourism and Environment, D.A.J. Wong, who left no doubt of his government's determination to change the way of life of the Penan people, writing:

> In the last 40 years, the natives of Savanah, some of whom were then still living primitively, have been brought into the mainstream of our civilisation. The nomadic Penans, however, have always posed a problem as they choose to live as wandering nomads in the jungles. They pose a dilemma to the Government. . . . Efforts to get together the nomadic Penans and persuade them to settle down is almost an impossible task. Nevertheless the Government has and is still doing its best to do so.
>
> (Wong 1988, p.16)

Such attitudes were the conventional wisdom twenty years ago, almost universally held by governments and even well-meaning Western liberals practically worldwide. Now they have the unmistakable ring of bigotry shot through with vested interest and are becoming increasingly rare. Indeed, the Colombian government of President Virgilio Barco recently recognised Indian land rights to over 18 million hectares of the Colombian Amazon. Although these rights amount to less than full ownership of the land, and there remain problems with government/Indian relations elsewhere in Colombia, such recognition is undoubtedly a positive step forward.

There is also a growing international awareness that tribal cultures have much to teach the industrialised and industrialising world about sustainable living in fragile ecosystems which Western-style development only seems to know how to pollute and destroy. Moreover, these ecosystems, especially the tropical rainforest, are being increasingly recognised as crucial to a sustainable Western way of life. Chico Mendes, the rubber-tappers' leader assassinated by ranching interests in December 1988, was not himself a tribal person, but spoke for many forest-people including tribal peoples when he said:

> At first I thought I was fighting to save rubber trees. Then I thought I was fighting to save the Amazon rainforest. Now I realise I am fighting for humanity.
>
> (quoted in Gaia Foundation 1989, p.4)

Indigenous peoples are increasingly articulating this sort of message themselves, and speaking out in their own defence, and Survival has laid great emphasis on giving them the space to speak and be heard internationally. One of the most effective of their organisations is the Coordinadora de las Organisaciones Indigenas de la Cuenca Amazonia (COICA) which now has organisational representatives from Colombia, Brazil, Peru, Ecuador and Bolivia and speaks for over half of all Amazon Indians. The COICA President Evaristo Nugkuag Ikanan received a Right Livelihood Award in 1986, following a nomination by Survival. In his acceptance speech, Ikanan said:

> Our sons and daughters are children, our women and men are people, just like any other, our nations have their pride, their history, their heroes, their beliefs, their customs, just like any other. Now through our organisations, we are getting stronger, from village to village, from country to country. Our principal objective is to promote an independent response to the day to day problems faced by indigenous populations and defend our rights from our indigenous perspective without outside interference. Our principal line of action is the defence of our land and our resources, as well as our right to our own language, culture and education, to self-determination and political representation for the security of our people.
>
> (Ikanan 1986)

On his way back to South America, and with two colleagues from Colombia and Bolivia, Ikanan had a meeting with Barber Conable, President of the World Bank, in Washington, winning new assurances of the Bank's future sensitivity to Indian needs. Such meetings are a striking expression of indigenous peoples' new-found confidence. From being viewed as primitives due for quick integration or extinction, tribal peoples are increasingly seeing themselves and coming to be seen as guardians of priceless natural assets. More and more people would agree with Harrison Ngau, the young Kayan Indian who was organiser of the Sarawak office of Friends of the Earth, and is now a member of the Malaysian Parliament, when he said: 'All over the tropical rainforest world, the people who are the best defenders of this remaining precious life-support system are the forest peoples.' (message to the launch of the Forest Peoples' Fund, London, November 1989, quoted in Gaia Foundation 1989, p.1).

Seventh Generation Fund (USA)

Of all the world's major indigenous peoples, none has a longer record of persecution by white men than the North American Indians, treatment which has brought them to the very edge of cultural extinction. Yet here as elsewhere, the new perception among North American Indians and many other Americans is that the Indian traditional culture has much to offer all of North American society and should be enabled to play this role. Such is the prevailing attitude in the Indian organisation the Seventh Generation Fund (SGF).

SGF was created in 1977 by Indian community activists to provide funding and technical assistance directly to local Indian communities. The organisation was founded on the premise that Indians need to move beyond the rhetoric of sovereignty towards concrete efforts to rebuild the economic and political infrastructures of native communities.

SGF is named after the old Iroquois Indian custom of considering the impact of all major decisions (before they were taken) on the seventh generation, a far cry from the short-termism and high future discounting dominant in today's Western societies. SGF has prospered, both as a project development organisation in its own right, and as an intermediary between such projects and outside funding agencies. In 1987–8 SGF granted US$127,000 to thirty-seven projects and helped to generate and arrange US$227,000 in supplementary funds (Seventh Generation Fund 1988). Its budget in the same year was nearly US$1 million, employing fourteen staff in six offices throughout the US.

The projects of SGF have four principal foci:

- *Self-reliant economies* – efforts which demonstrate the economic viability of small-scale community development which utilises local renewable resources and human skills to provide basic goods and services and which are appropriate to Indian culture; redevelopment of self-reliant economies through food production, appropriate technology, and alternative energy use.
- *Indigenous ways of life* – efforts which restore indigenous forms of political organisation or which modify existing governments along traditional lines, or restore indigenous community systems and apply traditional thought to contemporary issues.
- *Land and natural resources* – efforts to reclaim and live on aboriginal lands, or to protect tribal lands, resources and sovereignty;

- *Native women* – efforts initiated by Indian women to promote the spiritual, cultural, and physical wellbeing of the native family and which are consistent with Indian community/national values and ethics.

The problems of realising such aspirations in the current Indian condition in the US and Canada are legion. Indians themselves are divided and the divisions are skilfully exploited by the relevant government agency, the Bureau of Indian Affairs. The attempted destruction of Indian culture has produced many social problems on Indian reservations – drugs, drink, violence – which the governing tribal councils all too often do little to combat. And, as in the white community, there are genuinely held differences of opinion among Indians about their most appropriate development path. Some favour assimilation. Others look forward to a conventional Leftist revolution. Others still, like the SGF, are advocating a third way based on respect for indigenous culture, self-determination for Indian communities and land rights commensurate with their history, on the basis of which a more equal relationship between the Indian and white populations could come into being. There is no doubt that this 'Third Way' approach is growing. It is too soon to say whether it will become the dominant North American Indian aspiration.

CONCLUSION

Just as HURIDOCS is developing a communications network for human rights groups, so there is in preparation under the auspices of the United Nations University a computer database, a 'World Guide of Ethnic Minorities'. The work is being conducted by Rodolfo Stavenhagen at El Colegio de Mexico in Mexico City. Sensitively employed, such a database could do much to overcome the isolation of relatively small ethnic groups and create a sense of supra-national solidarity. Two hundred million is a sizeable world community, larger than the great majority of nation states. If indigenous peoples, supported by such organisations as Survival International, can be effective and united across international barriers, as is now starting to happen, then they stand a good chance of emerging from this traumatic, near-extinctive period of their history with better prospects than perhaps at any time since the Western colonialists and adventurers first arrived to destroy their communities and ways of life.

As has already been stressed, it is vitally important for non-tribal peoples that indigenous cultures survive. Native peoples have so much to teach the rest of the world, not just about rainforests and conservation but about the very meaning of life beyond 'development':

Why do people think that tribal peoples' cultures are inferior?

In the Third World there is a truly startling paradox when you compare these people with most others. Take Amazonia, for example, where traditional Indians live well in comfortable dwellings – warm at night and cool in the day. They eat well – a varied and healthy diet. They live in a close community where loneliness is unknown. And they do it all on 3 or 4 hours of work a day, or less, and have plenty of time for playing with their children, for contemplating philosophy, cosmology and religion, and for externalising whatever answers they find through profound rituals which make many of our own seem shallow and meaningless.

Compare this life with the lot of the Third World poor who supposedly benefit from 'civilisation' and who are by and large growing poorer daily in spite of the billions in aid. Their children are working 15 to 16 hour days. They are badly nourished. Serious disease is rife and western technological medicine largely unobtainable. Infant mortality high, life expectancy low, alcohol and drug abuse common – social breakdown is often the norm in the shanty towns where life comes and goes on the cheap.

The resource extraction, the dam building, most of the so called 'development' going on in these countries benefits neither the tribal peoples whose lands are destroyed in the process, nor the vast majority of the nation's citizens, the poor and needy. The ones who profit are, of course, the wealthy – and the governments are *always* wealthy – and the foreign companies.

(Corry 1989)

It is to consideration of further alternatives to this 'development' that we now turn.

5

CONTRASTS IN
DEVELOPMENT

The next two chapters seek to go beyond the critique of conventional development articulated earlier by explaining in some detail both a classic example of such development, the Narmada Valley Project (NVP), and the activities of many organisations round the world seeking to realise what has come to be called 'Another Development'. The Narmada dams, and the struggle against them, provide probably the single clearest example of the main contesting views in the development debate today.

DAMMING THE NARMADA

Big dams have long been a prestige symbol of industrialisation and development. India's first post-independence Prime Minister Pandit Nehru well-expressed their numinous quality in this context when he called them the new 'temples' of a modernising India.

As the development dream has become increasingly perceived as a nightmare, so too the reality of the big dam as opposed to its mythology has come under intensive scrutiny focusing on problems such as those identified by Timberlake (1985), including the displacement of people, salinisation of land, water-borne diseases, increasingly unequal distribution of wealth and 'disruptions to fisheries, loss of forests and wildlife, siltation due to erosion, excessive water losses due to evaporation and even an increase in earth tremors.' (p.85). The most formidable critics of the big dam are probably the team from *The Ecologist* magazine, Edward Goldsmith and Nicholas Hildyard, whose two-volume major study *The Social and Environmental Effects of Large Dams* appeared in 1984 and 1986. Reviewing volume 1 in *The New Civil Engineer*, Appleton (1984, p.23) wrote: 'The authors have collected a unique dossier of the effects of big water projects, and the picture they paint justifies the book's own description of "massive ecological destruction, of social upheaval, disease and impoverishment"'. Since these volumes *The Ecologist* has returned repeatedly to the dam theme, publishing such articles as S.K. Roy (1987) 'The Bodhghat Project and the World Bank', B.S. Cox (1987), 'Thailand's Nam Choam Dam: a disaster in the making', N. Jhaveri (1988) 'The Three Gorges Debacle', and L. Lohmann (1990b) 'Remaking the Mekong'. A quotation from the Jhaveri article conveys the general message of these articles. 'Its ecological and social impact will be devastating, with over 750,000 people due to be resettled and a host of unique habitats flooded. The project is also likely to increase salinisation, destroy valuable fisheries and increase pollution.' (p.56). Goldsmith and Hildyard's recommendation to decision-makers in the light of these sorts of effects is blunt: to cut off funds from all large-scale development schemes that they may plan to finance, or are involved in financing, regardless of how advanced those schemes may be. It is a recommendation which has cut no ice at all with the principal promoters of the Narmada Valley Project.

The Narmada is India's largest western-flowing river and its fifth largest overall. Its 1,312-kilometre course to the Arabian Sea winds through hills, forests, agricultural plains and rocky gorges in a series of falls, rapids and slack waters. Some 21 million people live in its basins, including a sizeable tribal population. The river is of great cultural and religious importance to the local people as well as, obviously, fulfilling important economic and ecological functions.

According to the World Bank (1989b), the Narmada river 'is one of (India's) least used – water utilisation is currently about 4% and tons of water effectively are wasted every day when it could be put to use for the benefit of the region' (p.3). This is not a new perception. There have been proposals to dam the Narmada at least since 1947 but these were stalled for over thirty years because of failure between the three riparian states involved – Gujarat, Madhya Pradesh (MP) and Maharashtra – to agree on how the costs and benefits should be shared. This issue was finally adjudicated by the Narmada Water River Tribunal in 1979 and planning for NVP got under way in earnest.

In total NVP is the biggest river project ever planned in the world. It comprises two very large dams – the Sardar Sarovar Project (SSP) and the Narmada Sagar Project (NSP) – twenty-eight other major dams, two of which have already been built (the Tawa and Barna dams) and some 3,000 medium and minor 'water projects'. Clearance for the building of SSP was given in 1987 and of NSP in 1988. The World Bank has agreed a US$450 million loan for SSP, part of which has already been disbursed, and is considering support for NSP. It has written:

> Taken together the two projects would help to irrigate some 2.2 million ha. of land in MP and Gujarat, and another 70,000 in Rajasthan. Of the 4.5 to 5 million people cultivating land to be irrigated, many are desperately poor. . . . Both projects will produce electric power, an essential ingredient for development of industries and agriculture, (which) has been extremely scarce, and shortages have hampered industrial production and employment in India substantially. The projects are also designed to moderate floods.
>
> (World Bank 1989a, p.3)

Pisciculture is another proposed benefit of SSP, as is the provision of drinking water over an enormous area. Indeed, Mr P.A. Raj, Vice-Chairman and Managing Director of the SSP construction company, has claimed (in a publication rather ingenuously entitled *Facts* [1989]) that: '4,720 villages and 131 urban centres in Gujarat State would get a permanent solution as regards the drinking water supply needs' (p.5).

This is just one of the 'facts' which are hotly disputed by the vigorous campaigns against SSP and NSP which have sprung up over the last few years. These campaigns did not originate as anti-dam campaigns but rather as activist groups seeking to ensure that those displaced by the dams were properly compensated and that the environment was safeguarded. It became apparent to these groups that neither of these conditions was going to be adequately met, and that other aspects of the whole project gave great cause for concern. The groups then made common cause against the dams as such and have coalesced into one of the most powerful social movements ever to emerge in post-independence India. A recent show of strength was a 60,000 strong demonstration in September 1989, in the small town of Harsud in MP, which is due to be submerged by NSP. Thousands of the demonstrators pledged themselves never to move from their homes.

In the campaign against the SSP villagers are increasingly refusing to co-operate with officials seeking their relocation and on several occasions brought some construction work to a halt by non-violent actions. There is little doubt that these sorts of confrontation will increase unless the outstanding problems are solved to general satisfaction, wherein lies the greatest problem of all: that the knottiest problems actually appear to be insoluble. The following analysis indicates the main points of contention.

Resettlement and rehabilitation (R&R)

SSP and NSP between them will displace or seriously disrupt the lives of at least 200,000 people, referred to officially as project affected persons (PAPs).

The World Bank (1989a) 'now requires that each project involving involuntary displacement have a comprehensive, detailed and feasible plan for resettlement and rehabilitation of the displaced population before it gives its approval for a loan.' (p.3). This requirement, and the pressure from activist groups which initiated it, won from the government of Gujarat (GOG) terms for R&R that are considerably more liberal than elsewhere in India.

However, terms are one thing, their implementation is another. The report by Thayer Scudder in May 1989, consultant to the R&R sub-group of the World Bank's Reappraisal Mission for SSP, concluded:

As documented in detail in this report the GOG has attempted neither to implement expeditiously its new policies nor to correct deficiencies (in the resettlement) of those relocated prior to December 1987 (when the new liberal policies began to be announced). During April 1989 I made a special effort to search out examples of 'good will' on the part of GOG towards its oustees in regard to a number of long-standing trouble points which the government, at small financial cost and a significant benefit to oustees, could easily have relieved. I found no such cases.

(quoted in Engineer 1989, p.93)

The GOG R&R terms are often presented by protagonists of the dams as if they applied to all the oustees, but in fact they are only relevant for Gujarat residents who comprise only 20 per cent of the SSP's PAPs and none of the NSP's (thus less than 10 per cent of the joint total). The other states have their own R&R policies which, according to Engineer (1989) 'are much worse in decisive respects' (p.94). He also quotes the January 1989 report by the National

Institute of Construction Management and Research (NICMAR), appointed by the central government to monitor compliance with the stipulations of the Narmada Water Disputes Tribunal and the World Bank: 'As a central monitoring and evaluation agency, we tried to locate if there was any detailed implementation plan for the R&R work. We have not come across any such plan, central or interstate or even at the level of each involved state' (p.94).

Given the lack of such plans, it is not surprising that such relocations as have already occurred have been disastrous for the oustees. Reports by the Centre for Social Studies, Surat, and the Tata Institute for Social Studies, Bombay, (reported in Amte 1989 and Kalpavriksh 1988) indicate widespread problems with the resettlement sites, including severe shortages of fuelwood, fodder and employment opportunities, fragmentation of families, conflicts with host populations, inferior land, shortage of water, health problems, and late or deficient compensation. The experience of the recently completed Tawa dam is also not encouraging. According to Alvares (1989) it has 'produced a veritable nation of beggars and moving migrants who have not found a piece of land to settle on ever since' (p.3).

Facts either denies or brushes off these problems with the confident assertion 'all the outstanding issues of PAPs will be resolved shortly' (p.12). *Business India*'s editor, Rusi Engineer, comes to a very different conclusion:

> It is clear that as of today there is nothing even remotely resembling a comprehensive and detailed action plan for R&R and that acceptable policies applicable to all oustees do not exist. . . . Nor can anyone realistically claim that the requisite policies, plans and implementing machinery will be brought into being in the short time before the remaining displacement is due to begin. The conclusion is therefore inescapable: if construction of the dam is allowed to proceed there is little doubt that a large number of poor people will be pauperised and condemned to a lifetime of destitution . . . futile wandering in search of work, toiling in the most exploitative conditions, begging on city streets and scrounging around in rubbish dumps, shunned by all and in the end reduced to such dire straits as to be forced to sell their wives and daughters. Such is the sacrifice being demanded at the altar of 'progress'.
>
> (op.cit. p.94)

Even more unequivocally, Engineer concludes later in the article that 'Satisfactory resettlement and rehabilitation of the oustees is just not possible' (p.95). In May 1990 the Japanese government, which had planned to loan US$150 million to SSP withdrew from the project with only US$20 million committed, because of inadequate R&R arrangements and mounting opposition to SSP in India.

Environmental impacts

The reservoirs of SSP and NSP will together submerge over 100,000 hectares of land, including 50,000 hectares of forests, much of it in NSP's case of absolutely prime importance in terms of timber, wildlife and genetic resources, of which there has not even yet been a definitive study. In 1987 India's Department of Environment and Forests estimated the environmental losses due to submergence by these dams at nearly Rs 40,000 crores (1 crore = 10 million) (DEF 1987), a sum equal to three times the total estimated cost of constructing SSP. According to Kalpavriksh: 'This of course has thrown the whole cost–benefit ratio of these dams out of gear; indeed, attempts by the project authorities to counter this border on the ludicrous' (op. cit., p.4).

Other environmental concerns raised by Narmada critics include accelerated siltation of the dams due to their degraded catchment areas; health impacts from water-borne diseases, waterlogging and salinisation of farmland, the effects on downstream ecosystems and the impact on backwaters. Some of these effects are potentially very serious. Siltation could significantly reduce both the life and performance of the dam. Waterlogging could have a similar effect on its irrigation benefits. The Indian Institute of Science has estimated that 40 per cent of NSP's command area is prone to waterlogging (quoted in Amte 1989 and Kalpavriksh 1988). Most worryingly, both the Tawa and Barna dams in the area have produced significant waterlogging, to the extent that crop yields in one district actually declined after irrigation by Tawa water.

India's Department of the Environment and Forests (DEF) has laid down conditions for catchment area treatment, compensatory afforestation (a misnomer because no human plantation can compensate for a mature tropical forest) and command area development. As with the GOG's liberal R&R policies, much is made by the

NVP promoters of these conditions as evidence of the scheme's overall environmental responsibility but, again as with R&R, the critics contend that the conditions are more honoured in the breach than the observance. Amte (1989) writes despairingly of the Environmental Sub-Group (ESG) of the Narmada Control Authority, which is supposed to monitor compliance with the DEF conditions:

> A careful examination of the minutes of the meetings of the ESG shows that it has played no effective role in ensuring the implementation of any of these conditions. In the five meetings of the ESG held so far its members appear to be helpless observers as written conditions, explicit deadlines and even desperate oral warnings have been brazenly ignored.
>
> (p.24)

Amte's despair, however, seems to have been premature. In September 1990 the ESG deemed environmental clearance for SSP to have lapsed, because of failure by the project authorities to meet the Ministry conditions. As the Planning Commission's clearance of the project was conditional on environmental clearance, anti-Narmada activists claim that the project as a whole has become illegal (Kane 1990).

The World Bank

The World Bank decided in 1985 to fund SSP through a loan of US$450 million, one of the commitments which earned it the charge of 'bankrolling disasters' from an increasingly outspoken and influential international consortium of environmental and tribal-people pressure groups, which has monitored its work. At that time the Bank paid little attention to such issues but pressure group impact was such that in 1987 the Bank's President signalled a major change in his organisation's policy. If the Bank had hitherto been part of the problem in these areas, Conable declared, it would henceforth be part of the solution (Conable 1987).

The decision by the World Bank to continue funding SSP despite the obvious fact that its conditions on R&R and environmental protection are not being complied with has done much to discredit this declaration in the eyes of the pressure groups, many of which had given an initial welcome to the Bank's apparent change of heart. Most infuriating was the Bank's decision to extend for a third time to June 1990 its deadline for compliance with its R&R stipulations, despite the very critical report of its own R&R Mission in May 1989 (as discussed earlier). This drew the following comment from the Narmada Campaign Newsletter of August 1989:

> The moral bankruptcy of the World Bank (WB) has been revealed in its recent decision to continue funding SSP, extending its own deadline by another year, in spite of the scathing indictment of its own reappraisal mission. . . . (This decision) exposes the continuing hypocrisy of the World Bank. . . . The various ecological issues relating to the dams also continue to receive short shrift, both from the Project authorities and the World Bank. . . . It is clear that the Narmada dams are being pushed ahead with little regard for the ecological destruction they cause.
>
> (NBA 1989, p.5)

The campaigning groups' conclusion was only confirmed when in June 1990 the Bank decided to continue extending credits for another year despite detailed documentation from the Narmada area as to how their previous stipulations were very far from being met by the project authorities (Patkar 1990).

The campaigners' attitude is not confined to the local campaign against NVP. The US-based Environmental Defense Fund (EDF – see also Chapter 6) has been following the issue closely for four years. In recent Congressional evidence its staff attorney Lori Udall opined: 'At this point in time most NGOs familiar with the project (SSP) cannot understand the Bank's continued funding of a project so riddled by social unrest, poor planning, misinformation and outright negligence in environmental, resettlement and economic issues' (Udall 1989b, p.2).

Nor can it be claimed that SSP is an unfortunate blot on an otherwise respectable World Bank record on these issues. Udall relates:

> *Worldwide, out of approximately 56 projects that the World Bank is financing involving forced resettlement, the Bank cannot document one single case where the population that has been resettled is better off than before or has reached the standard of living which they had before.*
>
> (Udall 1989a, p.4, original emphasis)

Bruce Rich, also with EDF, tells the same dismal story: 'Yet as NGOs in the North and South have encountered more and more Bank-financed ecological debacles, disillusionment with the Bank's environmental reform initiatives has grown' (Rich 1990a, p.307).

In researching the Narmada issue I wrote to the Bank asking for information. I received three pieces of documentation: the highly contentious *Facts* put out by the construction company and two short World Bank papers, listed in the Bibliography as 1989a and 1989b. Neither document mentions that the very significant problems discussed here are still outstanding. 1989a appears to renege on the Bank's own R&R conditions when it says: 'The Bank will make every effort, in working with concerned state governments, to ensure that an alternative livelihood is provided (to oustees), preferably one based on agricultural land' (p.4). Such a sentence makes light of the fact that it was supposed to be a condition of the Bank's involvement in the project that such an alternative and equivalent agricultural livelihood for oustees was *guaranteed*. 1989b totally fails to mention the scathing R&R report received from the Bank's own mission some four months earlier and seems resigned to the oustees' sacrifice:

> (The SSP) can be accomplished by *only* dislocating 70,000 people whose current livelihood is largely below minimum subsistence levels in an area where the ecology/environment has already degraded to near or below an acceptable bottom line. . . . It is something of a miracle that only 70,000 persons would be physically affected by the creation of a 210 km long reservoir in a nation of 800 million people in one of its largest and most important river basins.
>
> (pp.3,4; emphasis added)

1989b is also guilty of conveying false impressions when it says: 'Non-governmental groups in the project area are regularly contacted by Bank review missions and, in fact, are involved operationally in aspects of the project such as monitoring and evaluation and land purchases for displaced persons' (p.2). In fact the groups with which the Bank is working in this way are not representative of those due to be displaced. On 17 July 1989 the coalition of anti-Narmada groups, including representatives of oustee villages in Gujarat, Maharashtra and Madhya Pradesh, demonstrated against World Bank involvement in SSP outside its office in Delhi and issued a press statement announcing:

Representatives of the communities to be displaced by the Project have today asserted that they will not allow any World Bank or Government official to enter their villages for any work related to the dams. The struggle opposing the dams is going to be expanded and intensified at all levels.

(quoted in NBA 1989, p.7)

The World ? ank has a lot to learn in terms of even-handed public relations and good faith if it wants its new self-professed image of environmental and human concern to be taken seriously by an increasingly aware public and politicians. As for future decisions, this image will clearly have no credibility at all if the Bank gives financial backing to NSP without prior in-depth studies of its impacts and root-and-branch redesign of the project to make those impacts ecologically and socially acceptable. Perhaps this message, at least, has penetrated the Bank, for NSP has now been dropped from the World Bank's list of pipeline projects in its monthly operational summary.

I sent to everyone who features largely in this book a copy of the section on their work, with a request for comments. Thomas Blinkhorn of the World Bank India Department's Agriculture Division replied that my account of the Narmada issue was 'so replete with error and misleading judgements that it is virtually impossible to proceed without rewriting the entire (section).' (personal communication 9 October 1990). He therefore sent me four pages expressing the World Bank's view of the matter with the request that I substitute this for my text or publish it as an addendum.

Because the new World Bank statement adds little to the two documents (World Bank 1989a, b) already quoted, simply re-presenting as facts the project's supposed benefits, which have been so hotly contested, there is little reason to reproduce it here. However, a letter to the World Bank at the address given in Appendix 2 will doubtless yield a copy. I suggest that those who want a far more detailed critique of the Narmada dams than the brief and objective survey I have been able to give here write either to Baba Amte for a copy of *The Case Against Narmada and the Alternative Perspective* (Amte 1990) or to Lori Udall, Environmental Defense Fund, for the section on Narmada in the citation Udall 1990 in the Bibliography. Again, addresses are in Appendix 2.

Alternative futures

The options of any government of India that decided to review the Narmada decisions are well-defined. Cancelling the dams would take enormous political will and courage, in view of the funds already committed and the considerable popular support for them, especially in Gujarat, based on the highly doubtful claims by the authorities that the dams comprise 'a real drought-proofing Project and hence the life-line of Gujarat' which 'will be a permanent solution for drinking water supply problems of water-hungry regions of Saurashtra and Kachch.' (Raj 1989, pp.3–5). Not cancelling them will probably condemn over 200,000 people to lives of utter misery and compound India's ecological problems for uncertain benefits. It will also cause much prior bloodshed through civil strife, given the new militancy of the oustees in their determination not to be sacrificed quietly, and the declared readiness of the Chief Minister of Gujarat to use troops against the protesters if necessary.

A notable fact of the World Bank documents reviewed here is their total dismissal of alternative ways of resolving the area's water problems. This is in striking contrast to the critics' recommendations. Amte (1989), for instance, recommends a broad strategy including fuller utilisation of existing installed irrigation capacity (58 per cent of this capacity in Gujarat and 70 per cent in MP is apparently unused), more economical use of water, especially by industry, and a strategy to improve ecological balance, through afforestation and agroforestry, improved dry farming technology, erosion control measures and small-scale water harvesting. The big dam choice completely forecloses these alternatives because its enormous cost is bound to absorb all available resources (in a situation reminiscent of the nuclear power/alternative energy debates).

Engineer (1989) elaborates on the last of Amte's recommendations, improving ecological balance, in a way that looks forward to future chapters of this book, so this section will be brought to a close with a substantial quotation from this source.

> It is the destruction of the once rich pastures and forests of these areas which is the real cause of their acute water crisis today. . . . Once this connection is seen, restoring vegetational cover to the land becomes the first priority. . . . This is not romantic and fanciful 'back to nature' nonsense. Village level

efforts in several drought-prone areas of the country have conclusively demonstrated the truly amazing results that are possible with a programme of small watershed development and ecological restoration, rainwater harvesting and small-scale storage and increased efficiency of water utilisation.

(p.99)

Engineer then gives the example of Ralegan Shindi, a village in Maharashtra

not very different from large parts of Gujarat to be served by SSP. Before the watershed development programme was initiated just a few years ago, the village faced acute drinking water shortage, yields of jowar and bajra were very low, and more than half the food requirements of the village had to be purchased from outside. After the watershed programme, drinking water shortage has been totally eliminated, irrigated area increased twelvefold, the cropped area as well as yields have more than doubled and the village is now a net exporter of foodgrains. All this has been achieved in just a few years, with limited means, without using any exogenous water from any large system – and results are expected to be even better in the coming years.

(p.99)

ANOTHER DEVELOPMENT

The essential components of a radically different form of development to that promoted by the Brandt Reports and, to a lesser extent, the Brundtland Report and epitomised by the Narmada dams, were articulated nearly fifteen years ago, in a pioneering publication from the Dag Hammarskjöld Foundation (1975) entitled *What Now? Another Development*. These components of 'Another Development' were expressed in a later Dag Hammarskjöld (1977) publication as five in number. Another Development would be:

- *Need-oriented*, that is being geared to meeting human needs, both material and non-material.
- *Endogenous*, that is, stemming from the heart of each society, which defines in sovereignty its values and the vision of its future.
- *Self-reliant*, that is, implying that each society relies

99

primarily on its own strength and resources in terms of its members' energies and its natural and cultural environment.

- *Ecologically-sound*, that is, utilising rationally the resources of the biosphere in full awareness of the potential of local ecosystems as well as the global and local outer limits imposed on the present and future generations.
- *Based on structural transformations*, required, more often than not, in social relations, in economic activities and in their spatial distribution, as well as in the power-structure.

(p.10)

All five of these components have been further developed, explored and refined since the *What Now?* report first appeared, but none have been made redundant, nor have any others been added. Another Development has emerged as a clear and coherent system of developmental analysis to contrast with the top-down, finance-oriented economism of conventional development strategies.

A prime example of this approach in practice, clearly showing the success and achievements that are possible, but also the enormous problems of working for Another Development in a very hostile external environment, is the Sarvodaya Shramadana Movement of Sri Lanka.

Sarvodaya Shramadana Movement (SSM) (Sri Lanka)

SSM was founded in 1958 by A.T. Ariyaratne when he was a 26-year-old teacher. The name literally means 'Awakening of all by voluntarily sharing people's resources, especially their time, thoughts and efforts', with the movement's most prominent activity being the Shramadana Camp in which the whole village gives its labour in order to accomplish something of collective value.

By 1985 SSM had become probably the best-known development movement in the world, said to involve in 1987, directly or indirectly, some three million of Sri Lanka's fifteen million people, with 7,000 full- or part-time employees.

Out of a total of 23,000 villages in Sri Lanka the Movement is active in 8000. In each village a number of youths from the village itself, who have undergone training in different skills of village reconstruction, work as full time workers. They inspire, learn from, educate, organise and work with men, women and children in their villages on a programme of village self-development carved out by the village community itself.

There are over 30,000 trained village youths working in their own communities improving the quality of life of their people. The sectors in which they work include nutrition, health, education, housing, water supply and sanitation, irrigation, agriculture, communication, savings and credit, rural industries and marketing, legal aid, institutional building and spirituo-moral development. In short they are participant-beneficiaries of an integrated rural awakening programme.

(Ariyaratne 1985, p.i)

A balanced development, where material as well as non-material needs (such as, spiritual, social, cultural, etc.) are satisfied, is Sarvodaya's objective. For this purpose, Sarvodaya has identified Ten Basic Human Needs as follows:

1. A clean and beautiful environment
2. A clean and adequate supply of water
3. Minimum clothing requirements
4. A balanced diet
5. A simple house to live in
6. Basic health care
7. Simple communication facilities
8. Minimum energy requirements
9. Total education
10. Cultural and spiritual needs

The satisfaction of these ten basic needs from below is the strategy of Sarvodaya. The final satisfaction of these basic needs in all, leads to the awakening of all. While the basic needs of all are satisfied, this balanced development also acts as a constraint on the development of non-basic needs of a few. Therefore, a kind of social control is established on the rich and the affluent not to increase their inordinate needs (conspicuous consumption) at the expense of those less fortunate than themselves.

(ibid. p.5)

Sarvodaya villages go through five stages in their development process. First: An Initiation and Psychological Infrastructure Development Stage; Second: A Social Infrastructure and Training Stage; Third: Basic Needs Satisfaction and

101

Institutional Incorporation Stage; Fourth: Income, Employment Generation and Self-financing Stage; and Fifth: Sharing (with neighbouring villages) Stage.

The village communities go through these five stages basing their programmes on the three principles of Self-reliance, Community Participation and Planned Action. To assist them in their self-development efforts a national organisation known as the Sarvodaya Shramadana Association (Incorporated) has been established. The Association has established in every district of the country a district office and a district level Development Education Institute. At the sub-district level also there are Divisional Sarvodaya Offices and Divisional Sarvodaya Development Education Centres established. Altogether there are over three hundred such institutions in the country manned by a full-time trained staff. At one time approximately 2400 trainees from villages undergo residential training at these centres in skills needed for village development.

(ibid. p.ii)

Another Sarvodaya publication (SSM undated) expresses SSM's purpose and activity thus:

Villagers wishing to be assisted in the awakening of their village by Sarvodaya are visited by some of the Movement's staff members and together they identify the biggest or most pressing 'felt need' of the village that could be met by physical labour. This may be the desilting and restoration of an ancient irrigation tank (water reservoir), the construction of a new tank, the cutting of a new road or the rehabilitation of an existing road, or a similar manual work project. Sarvodaya then helps the village to organise a big Shramadana ('voluntarily working together') camp in order to tackle this project, using the labour of the local villagers, reinforced by the villagers from other villages already involved in the Movement, and by Sarvodaya volunteers from towns and cities.

A strict code of self-discipline is followed in these Shramadana camps with six to eight hours of each day devoted to physical work, and three to four hours to education through dialogue, song and dance. The purpose of these Shramadana camps is to catch the attention of the villagers to demonstrate to them that some of the solutions to their problems lie in their own hands (literally and figuratively), and to prepare the real process of awakening which begins even while the camp is in progress.

Beginning during the Shramadana camp, various village organisations, a children's group, a young people's group, a women's group and a farmers' group are created and then steadily developed. These organisations are the basis from which the village can begin to develop itself. Through these organisations, villagers can discuss with each other and then, if necessary with the representatives of the village council, their problems, needs and wishes, and together they can take whatever action is required. Very often, they find that by planning and working as a group they can meet village needs which, although simple, had remained unmet for years, for example, rehabilitating village roads and irrigation facilities.

These groups, with the help of Sarvodaya staff members, also spend time discussing the Sarvodaya philosophy – a Gandhian philosophy which incorporates elements of Buddha-Asoka tradition (Buddhism is the predominant religion in Sri Lanka) and other major religions, and pre-colonial Sri Lanka rural culture. This philosophy stresses four basic values – loving kindness, compassionate action, unselfish joy and equanimity supplemented by sharing, pleasant speech, constructive action and equality. The major long term goal is non-violent social change which will awaken everyone to live by these values.

When a village has developed sufficient experience in self-development activities based on self-reliance, community participation and planned actions, it forms itself into an independent Sarvodaya Shramadana Society and gets registered as an incorporated body under the Registrar of Societies of the Government of Sri Lanka. An Executive Committee of 25 elected members runs the affairs of the village society thereafter. In this committee three children between seven and fourteen years, three youths between fifteen and twenty eight and three women are also elected as full members. Such legal status obtained by the village society enables it to possess land and property, have access to bank loans, start individual and group economic enterprises, all of which help them improve their quality of life.

The villages are encouraged to work in a cluster of five for mutual help. A full-time Sarvodaya volunteer helps each cluster in a variety of ways and links it with a Sarvodaya Divisional Centre which organises training and other inputs needed for them.

In every administrative district in Sri Lanka there is at least one Sarvodaya Development Education Institute. There are 60 such Institutes at district level providing facilities for every kind of rural development training for village youths and leader's-training courses ranging from two weeks to two years and including training in both leadership and vocational skills.

Sarvodaya Headquarters which coordinates and channels support to these re-awakening villages is situated 20 kilometres (12 miles) south of Colombo. There are eight compounds located in 12 acres of land housing different kinds of programmes including a nutrition home for orphaned children, rehabilitation and occupational training for the physically and socially disabled, residential school for the deaf children and training centre for pre-school teachers, health-care workers, community leaders and so on.

(SSM undated, publicity leaflet)

Sarvodaya's programmes are carefully structured into separate, largely autonomous organisations. The key Lifeline Programme, involving the Shramadana camps' training courses, and the basic village development work, acts through twenty-three district centres, 225 divisional centres, 900 gramadana centres and 4,500 villages. Partial programmes, including relief and rehabilitation in the North and East, involve another 3,500 villages. It is estimated that in 1988–9 participation in the Shramadana camps amounted to about 415,000 person-days, cost about Rs4.6 million, of which over half came from the communities themselves, and produced value of the order of Rs17.4 million, quite apart from all the social and organisational value resulting from the ongoing groups: of the Lifeline villages, 72 per cent have children's groups, 47 per cent youth groups, 71 per cent mothers' groups, 70 per cent of the villages have pre-schools, for the financing of which they are totally responsible. SSM does, however, provide training for the teachers (chosen by the village). A popular activity is the children's fairs at which both children and adults sell their produce, with children encouraged to subscribe to a children's saving scheme, in which over Rs11 million are now deposited in over 180,000 accounts. To support this 'fieldwork' the districts run libraries and organise a wide variety of courses. Training and seminars are also the main functions of the divisional and gramadana centres, involving over 130,000 people, the majority (64 per cent) of them women.

Sarvodaya's welfare work is now handled by the Suwa Setha Services Society (SSSS), which runs nineteen units and handles relief and rehabilitation work due to natural disorders (i.e. not violence), including two vocational training centres for the handicapped, and several children's homes.

SSM also has an autonomous women's movement, operates legal aid services and an anti-drugs campaign, runs several special training institutes and services and a number of special projects and programmes. SSM's relief and rehabilitation (R&R) work in the North and East is carried out by a special directorate, on the basis of a three-year plan for R&R with a budget of Rs169 million.

A major recent growth area in SSM is its Sarvodaya Economic Enterprises Development Services (SEEDS), which encourages savings, extends credit, gives technical advice, carries out management training and seeks to develop small- and medium-scale businesses operating with a Sarvodaya philosophy (i.e. not profit-maximising). Over 1988–9 all the areas showed a healthy growth and 37 per cent of SEEDS operating costs were recovered as income.

In total SSM now employs over 10,000 people. Its external resource requirements amount to about Rs400 million (US$10 million) per year.

SSM's size, approach, purpose, activity and achievement have been subject to intensive study and review in numerous publications from the prolific Ariyaratne himself, from independent evaluators from funding agencies and from many other analysts and students who have sought to understand this movement from a variety of different viewpoints. One independent evaluation in late 1984 concluded:

> Sarvodaya is a remarkable organisation, probably unique in the field of international development. It derives its strength from its responsiveness and its Shramadana approach to village development, from which all other programmes develop. Commitment on the part of its workers, both those who are paid and those who are not, is Sarvodaya's greatest asset, one that has been purchased through the success of

105

its approach and the success of its programme. This success can be measured in the integration and the cost-effectiveness of Sarvodaya's programmes in villages or projects where the complete range of Sarvodaya activities is present. But it can also be found among Sarvodaya activities still in fledgling state. All members of the Review Mission were unanimous in their admiration for what the Movement has accomplished with resc rces which are minimal in comparison with the achievements.

(Smillie *et al.* 1985, p.39)

Thus SSM has, one can be reasonably certain, benefited millions of people. It has also raised hopes in millions more that here at last might be a successful 'alternative' development strategy capable of going to scale and displacing the Western model of industrial development. It is in relation to this grandiose possibility that most of the criticisms of SSM have been made, on the following grounds:

Dependence on foreign funding

Smillie *et al.* (1985) note this: 'Supporters and detractors alike have pointed to the apparent inconsistency in Sarvodaya's commitment to developing self-reliance at the village-level and its own dependence on foreign agencies to sustain its infrastructure and programmes' (p.22). SSM generated some 17 per cent of its Rs50 million income in 1983–4. After noting SSM's strenuous efforts to increase its income generation, Smillie *et al.* conclude 'There can be no doubt that SSM is in considerable need of programme support, and that this is justified by development activity as well as its cost effectiveness' (p.35).

This attitude is borne out by a later study (McBride *et al.* 1988) conducted in August/September 1987, by when SSM's annual budget had risen to Rs108 million. The study notes:

Sarvodaya is performing well and has achieved a great deal in the review period (since 1984). Donors' confidence in Sarvodaya's ability to spend their money effectively in the interest of Sri Lanka's rural poor is not misplaced. . . . (SSM) can take pride in its ability to effectively use and account for donors' funds.

(McBride *et al.* 1988, pp.51–2)

106

Goulet's 1981 study largely accepts Ariyaratne's distinction between self-reliance and self-financing and concludes: 'What is certain, however, is that Sarvodaya is fully prepared to "go it alone" if it were ever required to sacrifice its self-respect in order to win financial cooperation' (p.27).

Vulnerability to Western-style development

There is a worry that SSM's essential philosophy and approach will not be able to withstand the onslaught in success, or failure, of Western industrial development. This is the basic question explored in Goulet's fascinating study, which fears that the Buddhist basis of SSM's whole activity will be terminally undermined, especially among the young, by Western materialism and consumerism, and that SSM had neither fully recognised this danger nor was doing enough to combat it.

Inadequacy of political analysis

It has been charged that SSM has failed to identify or seek to tackle the key development issues of class, ethnicity and politics generally, a failure which was bound to render its efforts ultimately superficial. SSM's holistic, non-confrontational determination to work with all at the village distinguishes SSM from the much smaller Sri Lankan NGO, the Participatory Institute for Development Alternatives which carefully targets and works only with the poorest people in the villages, from the perception that the elite in any situation will always manage to corner the lion's share of the benefits of any untargeted improvement in living standards. Sisira Navaratne (1988) expressed thus the more general point about SSM's political stance:

> Sarvodaya tries hard to be a non-political organisation. But in Sri Lanka's tense political climate it is very difficult to be impartial when the goal of the organization is to 'awaken the entire nation' like Sarvodaya says. Some Sri Lankan government policies have opened us up to a barrage of foreign products and values which are contradictory to our culture. Sarvodaya talks about reclaiming our traditional value system while the government removes thousands of peasants from their lands to make way for massive irrigation schemes.

How can Sarvodaya talk about restoring values when it doesn't take a stand against the government?

Sarvodaya has a very good relationship with the present government. There is no material aid but it does have the government's political support. It's because Sarvodaya has never confronted the government that people assume the movement takes the side of those in power. This perception has got in the way of solving some of our country's most serious problems, particularly the ethnic violence.

Sarvodaya has continued some relief activities in the North and East like other NGOs. But it has not been able to deal with the main ethnic issue since the militants don't see Sarvodaya as impartial. Rather they're seen as being allied with the Sinhalese. Both Sinhalese and Tamils had hoped that Sarvodaya could play a greater role as mediator. But that couldn't happen because Sarvodaya was unable to hold a dialogue with the militants. In fact militants killed the Sarvodaya leader in the North as an indication of how they felt about Sarvodaya's political preferences.

If Sarvodaya wants to build strength among the villagers, it has a duty to help people understand larger political issues and act on them.

(pp.3–4)

The McBride *et al.* evaluation had a rather different perception:

The Movement has a pervasive influence in Sri Lanka. Its sheer size is a contributory factor. Equally important in this troubled country, however, is its commitment to working with all religious and ethnic communities, eschewing partisan politics and working consistently for peace and unity.

(op.cit. p.1)

It should not at all be assumed that SSM has been inactive in the face of the Sri Lankan ethnic violence. Detlef Kantowsky (1988) notes:

Sarvodaya alone seems to many to be capable of neutralising the tensions in the country and providing constructive forms of co-existence. Such hopes are not just founded on the spectacular peace marches and actions of the movement in recent years; they are based above all on the results of the ever expanding grassroots work carried out by Sarvodaya since the end of the 1950s in the villages all over Sri Lanka. For its non-partisan and non-denominational approach, still adhered to today, Sarvodaya has admittedly also had to make sacrifices. In September 1986 a group of Tamil terrorists tortured to death their fellow countryman, K. Kadiramalai, the 33 year old Director of Sarvodaya projects in the Jaffna area, because he was not prepared to support violence. Another seven Sar odaya (Tamil) village workers met with their death within the past few months in 'reprisals' in the north and east of the island, against Tamil Tigers by the Indian forces in which they were also accidentally caught.

(p.1)

and

A mere 24 hours after the massacres of July 1983 when no-one else ventured out of doors, the Movement began to organise relief for the injured and to set up refugee camps. Tamil and Sinhalese helpers worked together in the respective groups so as to set an example of reconciliation amidst the general group hatred.

On 2nd October of the same year a 'Peace Conference' was organised by Sarvodaya in the 'Bandaranaike Memorial Hall', the largest public building in Colombo. Two thousand distinguished persons from all over the country and the leading representatives of all religious communities participated. They passed a 'People's Declaration for National Peace and Harmony' which was carried in full by all the mass media in Sinhala, Tamil and English. The eight day peace march from the 'Temple of the Tooth' in Kandy to the sacred 'Bodhi Tree' in Anuradhapura from 3–11 June 1987 was an attempt to create a 'critical mass of peacefulness' throughout the country. More than 230,000 people all told took part in this action, and in the sitting and walking meditations together practised creating the capacity for loving kindness in themselves and how to pass it on to their fellow countrymen.

(pp.10–11)

Ariyaratne himself wrote a remarkable letter to Sarvodaya workers on 7 October 1987 in which he puts the Sarvodaya case. The following quotations give a flavour of its appeal:

We are an organization which has accepted non-violence. Our

109

first loyalty is to the common masses. We have to prevent them from getting into situations of violence. When they are faced with violence we must go in for relief-action immediately, to bring them solace. Whoever gets injured or whoever is deprived of food and shelter and other basic amenities has to be helped. We have to help without finding out to which caste, race, religion or political party they belong, even as we have been doing in the past. Our Movement is well-respected throughout the country as a relief and rehabilitation organization. The first thing we should do at this time is this: Whenever you hear that some people are in trouble go to those places, organize the community and do whatever is possible to give them immediate relief.

We should not get into any conflict or enter into any controversy which is outside our developmental, relief, reha-bilitation and reconciliation work. We should not get involved in any of these conflicts. There are enough conflicts in the country already and therefore we should not be a party to those conflicts. On the contrary, we should be a democratic force trying to unite all the peoples in our country, in a non-violent way to fight for justice and for peaceful solutions to our problems.

We do not consider anyone as our enemy; nor should anybody look at us as his enemy. We know all too well that there are people who use their political power against the common man; people who are prone to bribery and corruption, people who do not allow the common man to enjoy the freedom, justice and security guaranteed under the law. We know those who economically exploit the people. Such people as they will be opposed to us, not openly but in indirect ways. They have worked against us in the past, do so even now and perhaps in the future as well. We cannot help that. But the people who believe in progressive democratic changes in the country and those who work for justice will accept us and work with us. Only when there is a social change will we be able to overcome these obstacles and vested interests. Until then we have to even sacrifice our lives and undergo all kinds of oppression ourselves. We should not lose heart; but strive to continue with our work with great courage against such forces.

(quoted in Kantowsky 1988, pp.15–17)

It has to be said that the failure of Sarvodaya, despite its pervasive national presence, to stem the tide of terrorism and bloodshed that has engulfed Sri Lanka since 1983 gives added tragic weight to the criticisms of Goulet and Navaratne. Despite achieving so much for so many, it appears that Sarvodaya never succeeded in finding an adequate cultural base for its development programmes, which could withstand the combined hostility of the cultural assaults from the West, the economic policy of the Sri Lankan government and the internal, especially ethnic, contradictions within Sri Lankan society. Sarvodaya's approach has been far more sophisticated than the bald economism of conventional development thinking which has contributed at least to some extent to Sri Lanka's current social breakdown. But even Sarvodaya can now be seen not to have been rooted enough in its total cultural environment to have neutralised the forces that subsequently tore Sri Lanka apart.

6

DEVELOPMENT BY
PEOPLE

The Sarvodaya Shramadana Movement is just one example of an explosion of popular organisations and activities across the world seeking, consciously or not, to put the Another Development approach into practice. These initiatives were spurred both by the obvious dysfunction in the world economy and by the growing realisation that governments could or would not find the remedies for that dysfunction. Only people's organisations which began to solve the problems directly would show what needed to and could be done and would begin to generate the political will and reform to get governments working actively for the people rather than, in too many cases, against them. This chapter describes a selection of such initiatives, seeking to identify their common concerns and characteristics.

Six S Association/NAAM Movement
(Burkina Faso)

Both these organisations were founded by Bernard Lédéa Ouedraogo, who was born in Haute Volta (now Burkina Faso) in 1930. He completed his secondary education in Burkina Faso and gained many diplomas before studying in France, gaining a doctorate from the Sorbonne in 1977.

After finishing school in 1950, Ouedraogo became a teacher and school director and then turned to agriculture, where his talents as a trainer led him to the top echelons of the civil service, which he then abandoned in 1966 to found the NAAM movement. He explains why:

I was responsible for the training of the rural extension

workers, and young farmers who had some kind of formal
schooling. It was also my responsibility to supervise the
'official' village groups organized by the government (and
not by the farmers themselves). I did my best to help these
groups, but I failed. So I tried to find out why I had not
succeeded. What had happened?

The rural extension workers would arrive in a village, and
the only concern of the officially organized farmers was to
take advantage of the donkeys, bullocks, carts, hoes, and
other materials we would make available to them. But there
was nothing else behind this demeaning form of assistance, no
vision, no global conception of development or of the rural
world, no doctrine or philosophy. There had been no prior
efforts at consciousness raising. It was normal that in such
a situation the farmers had but one concern: prime the State
'pump' for all it was worth and cheat the extension workers.

So we asked ourselves: 'Is there anything in the organization
of traditional Mossi society that resembles these village
groups?'. We undertook a thorough study of village social
organization; of the people's thinking, of their social and
economic structures. We discovered that, of the village
organizations we examined, it was the NAAM group – a
traditional village body composed of young people which
undertakes various activities – that had the most highly
developed cooperative characteristics. We decided we would
attempt to work with the NAAM structures.

(quoted in Pradervand 1989, p.36)

The result, according to Pradervand, was an 'adventure which is
unique in the whole of Africa' (p.35). The NAAM groups prospered,
despite all the usual problems and official opposition. By 1987 there
were over 2,700 NAAM groups in the Yatenga area of Burkina Faso,
with over 160,000 members.

The NAAMs represent a triumph of the idea of 'developing
without harming' (Ouedraogo), of culturally appropriate development. The NAAM is a form of development adapted to
local needs, created by the people themselves, which instead of
destroying traditional structures from the outside, slowly, like
leaven, transforms them from the inside.

Many development specialists feel that the lack of authentic
popular participation has been the main failure of development

113

in the past thirty years. 'Development' has been something that has been done for people, to people, sometimes despite them and even against their will, rarely with them. The founder of the NAAM movement has shown that one can create a form of development 'of the people, by the people and for the people', hence his striking formula: 'letting oneself be mastered by the grass-roots'. The experience also shows that development is as much a way of travelling as a precise destination, and that the destination will be determined to a great extent by the way one travels.

(Pradervand op.cit., pp.39–40)

The transformation of the traditional NAAM groups into modern social structures was a masterpiece of practical sociology by Ouedraogo. He gives four reasons for their success: dynamic local leadership and activity; maintenance of traditional values; proscription of any sort of social, ethnic, political or religious discrimination; training and motivation coming from within the group based on the principle: act on the basis of what people *are*, what they *know*, how they *live*, what they *do*, what they *know how to do* and what they *want*.

The activities of the NAAM groups are as broad as life itself: social, economic, cultural. They grow, build, manufacture, trade. A World Bank report in 1983 listed their 'major construction activities' as embodying ten warehouses, nine cereal banks, seven other workshop buildings, three dams and forty wells, with funding from French, Dutch and Canadian bilateral sources and other international sources, as well as generating their own income.

Unaided, the rate of growth of the NAAM movement was necessarily limited by the rate of mobilisation of the villagers' own resources. It was to accelerate this process by tapping external funds that Ouedraogo founded the Six S Association (Se Servir de la Saison Sèche en Savane et au Sahel – the Association for Self-Help during the Dry Season in the Savannahs and the Sahel) with the French development expert Bernard Lecomte, in 1976. He became its Executive Director in 1978. While NAAM is a people's movement, Six S is an NGO dedicated to removing three obstacles to peasant mobilisation, as described by Bernard Lecomte:

The first obstacle was the lack of know-how. The farmers simply did not have the necessary knowledge to face the unprecedented challenges of the drought situation. The second one was the lack of 'negotiators'. By that I mean farmers capable of negotiating projects with both the local administration and the village elders without whose consent nothing could be achieved. The third was the lack of funds to implement small projects. More and more villages were becoming active, starting peasant groups, but once they started to organize on a regional basis, they were lacking in the funds both to support investments in organization and to initiate larger projects.

(quoted in Pradervand op.cit., p.154)

Six S is also intended to overcome the serious problem of under-employment in the region during the dry season, as its name suggests.

The structure of Six S is a federation of peasant organisations like (and including) NAAM, from nine countries in the region: Burkina Faso, Senegal, Benin, Mali, Togo, Niger, Mauritania, Guinea-Bissau and the Gambia. It is very tightly structured with especially strict control of financial disbursements, but that control is, and is perceived to be, firmly located in the farmers' groups themselves. To quote Pradervand again:

The General Assembly is composed almost entirely of peasant representatives who still cultivate their own fields. An aid organisation run by the farmers themselves is unique in the field of development aid, and one of the reasons that farmers identify so totally with Six-S and trust it.

(p.153)

In early 1989 there were 4,000 village groups in Six S (2,600 of them NAAM groups) divided into eighty-nine zones, forty-six in Burkina Faso, thirty in Senegal, eight in Mali, two each in Togo and Niger and one in Mauritania. The other countries were just getting organised.

Six S employs 166 people. Its income in 1987–8 was over 700 million francs CFA (US$2.7 million). Part of this would be loaned to village groups, part granted. The system of 'flexible funding' which Six S has evolved is another distinctive feature. It is not 'project linked' but spent at the discretion of the recipient once the group has shown itself creative and responsible. This further cements the trust in the organisation. In Burkina Faso groups repaid 60 million francs CFA in 1986–7, compared with only 1.3 million francs CFA in 1980–1, which indicates greatly increased economic activity. Figures of output are scarce, but it seems that farmers invest their work to twice the value of the external funds they receive, and 62 per cent of their time in 1987–8 was spent on environmental protection.

The key question for Six S is whether it can achieve its goal of self-reliance. There is no doubt about its determination in this regard. Its detailed strategy specifically calls for the withdrawal of financial aid once the federations have sufficiently progressed, making them rely on normal channels of investment, while their emphasis on loans from the first income-generating stages reflects their stress on prudence and self-reliance. But undoubtedly the farmers' temptation to regard Six S as just another permanent donor of aid is great and the swift growth of Six S may not permit the 'Six S Spirit' of self-reliance to be adequately disseminated. For the present it is clear that Six S and the peasant federations which it links are probably the only sources of hope for the 3¾ million people of Burkina Faso (and others in the other countries) whom it has directly benefited, in a striking achievement in twelve years of operation.

Bangladesh Rural Advancement Committee (BRAC) (Bangladesh)

BRAC was founded by F.H. Abed in 1972 following the Liberation War, initially as a relief project for refugees returning from India but thereafter a development organisation seeking:

> to improve the plight of the rural poor . . . by developing their ability to mobilise, manage and control local and external resources themselves. It was felt that BRAC's programmes should not be determined by a rigid set of strategies, but rather they should respond flexibly to the needs identified by the people. The ultimate objective is to end the long-standing exploitative relationships that dominate rural life in Bangladesh.
>
> (BRAC 1988, pp.1–2)

BRAC's growth has been phenomenal and it is now the largest development NGO in Bangladesh, employing as of September 1990 4,200 people with a budget of Tk800 million (US$22 million) and a membership of 350,000 (60 per cent women) covering 210,000 households in 3,200 villages, organised in 5,800 groups.

BRAC focuses exclusively on the landless, mobilising them into co-operative groups, who then plan, initiate, manage and control collective activities that lead to self-reliance. The activities cover a wide range of areas, reflecting BRAC's belief that the complexity of the problems demands a simultaneous search for solutions in many different fields, including:

- *Functional education*, the key process which villagers are required to complete before groups can be formed. In 1990 43,000 villagers were attending this course in 1,800 centres. The course itself consists of sixty lessons taught in two one-hour classes per day, six days a week for three-and-a-half months. It has also been used by other NGOs and the government.
- *Non-formal primary education*, for unenrolled children or drop-outs, using a specially-developed curriculum focusing on basic literacy, numeracy, health and environment. By the end of 1990 126,900 children, 70 per cent girls, were taking this course in 4,025 schools.
- *Training*, especially in the areas of human and occupational skills development, the former comprising consciousness raising, leadership development, project planning and management and functional education teacher training, the latter imparting skills in poultry keeping, agriculture and a variety of trades.
- *Meetings and workshops* – groups hold weekly meetings and inter-group meetings for a variety of purposes. Higher level committees of group delegates meet monthly to discuss issues that cannot be solved locally, e.g. wage bargaining, protest action, access to government services.
- *Health* – since 1980 BRAC has reached 12 million or 85 per cent of Bangladesh's rural households with its simple oral rehydration therapy for child diarrhoea (responsible for 33 per cent of Bangladesh's infant mortality). This has now developed into BRAC's Child Survival Programme, consisting of a primary health care programme, continuation of the oral rehydration work and assistance to the government in its immunisation and vitamin A distribution work.
- *Para-legal service* – since 1986 villagers chosen by their group are trained to act as para-legal counsellors in such matters as land conflicts and registration, civil rights and unfair practices.
- *Generation of income and employment/credit support* – agriculture, irrigation, fish culture, poultry, livestock, bees and other

rural industries have all been promoted as income earners against which credit is given (only after one year of group conscientisation and mobilisation). So far Tk427 million has been lent, with a repayment rate of 96 per cent.

On the direct production side, BRAC runs a sizeable Women's Production Centre, and has opened six shops called Aurong to market their products and those of 300 other producer groups. These are now self-supporting and directly employ ninety-eight people, turning over Tk69 million in 1989, and expected to turn over Tk100 million in 1990.

BRAC also produces a monthly magazine and has three commercial projects, BRAC Printers, a Cold Storage Enterprise and Garments Industry to generate internal income for the organisation.

BRAC's future plans include:

- *BRAC Bank*: BRAC is planning to formalise its credit aspect into a bank to make it separate from the awareness-raising and developmental work. Once this is operational it will be very interesting to compare it with the Grameen Bank (see page 122).
- *Citizen's report on the environment*: BRAC has brought together a group of major Bangladesh NGO's to compile a report on the Indian model.
- *Tobacco to silk*: Concerned that smoking in Third World countries is on the increase, partly because of aggressive marketing by tobacco companies to compensate for declining sales in industrial countries, BRAC has carried out studies of the potential of transferring land under tobacco to silk production. The potential is excellent: there is a large untapped domestic and foreign market and sericulture is twice as labour intensive as tobacco. BRAC's next step is to establish a 1,000 acre pilot project with farmers who have been persuaded to switch.

Working Women's Forum (WWF) (India)

WWF was started with 800 members in 1978 by Ms Jaya Arunachalam who had been prominent in the Congress (I) Party but had resigned the year before in disillusion about its effectiveness in reaching the poor. WWF was founded as a grassroots union of poor women workers in the informal sector. By the end of 1984 it had 36,000 members; three years later it had 60,000. By 1990 its members numbered over 150,000.

The initial need identified by its members on which WWF immediately began to act was for small loans at reasonable interest (the only alternative, moneylenders, charged interest at 10 per cent per month and more). WWF first acted as an intermediary between its members and a government loan scheme, administered through the nationalised banks, but the inflexibility and inappropriateness of the banks' procedures and attitudes caused WWF, with an initial seed-grant from Appropriate Technology International, to set up its own Working Women's Cooperative Societies (WWCS) to issue its own credit.

Only working poor women can become members of WWF, those who almost invariably suffer from the combined oppression of marginal work status, class, caste, gender, physical weakness and isolation. They (and their children) are usually ruthlessly ill-treated and exploited by moneylenders and their employers, as well as suffering the social gender bias against women. It is an extraordinary achievement for such women to have organised themselves in large numbers effectively to combat this oppression and materially improve their quality of life.

The success of WWF undoubtedly derives from the way it locates itself in and springs from the lives of its members. Robert Chambers has identified its key organising characteristics as: putting poor women's priorities first; working only with the poor; promoting leadership from below; and exercising clout to get the poor their rights (Chambers 1985, p.17). The clout has come from both Jaya Arunachalam's high-level contacts and WWF's ability to get thousands of women out on the streets of Madras in pursuit of their rights. Such clout brings real practical results: only the most foolhardy policeman or petty official now harrasses WWF members for bribes, which was a commonplace occurrence before.

An evaluation of WWF in May 1990 by the Dutch government (Netherlands 1990), drew attention to its record of leadership by working-class women members:

> By any standard, national or international, the Working Women's Forum (India) is a remarkable and successful institution. In terms of numbers, the Forum has more women staff and members than probably any other non-government organization in India and all but a few non-government organizations in the World. Even more notably, almost all of the staff are from the same socio-economic class as the

membership: that is, poor working class women.

The organizational structure, which promotes leadership from the membership itself, is the unique and dominant feature of the Forum. The governing body or executive council of WWF, with the exception of its Founder-President, are all working class women.

Also, all the WWF staff are working class women with the exception of a few administrative staff hired for their writing or accounting skills. This organizational structure, with the possibility of mobility from target group to staff positions is the main strength of the Forum.

WWF now operates in eleven separate locations, having spread out from Madras to start groups elsewhere in Tamil Nadu and in Karnataka, Andhra Pradesh, and Uttar Pradesh. Recently WWF has started organising workers in the Lucknow city slums and to unionise migrant labour women and help their children in Kampur industrial area. Although the women in these locations are in different occupations, the WWF process has met with success in every case.

Credit remains the core WWF service to members, and by the end of 1990 113,000 WWF members had received loans totalling Rs39 million (US$16 million) through the WWCS, while 28,000 members had been lent Rs6.5 million through the nationalised banks. Undoubtedly its success across different occupational groups – fisher-women, traders and vendors, bidi-rollers [a bidi is a local cigarette], lace-makers, agricultural fieldworkers and embroidery workers – comes from its efficient, empowering delivery of this crucial input, but WWF is active in other areas too. It runs a Grassroot Health Care and Family Welfare Project, jointly now with ILO, UNFPA and the government of India, employing 400 slum and rural landless women among the 90,000 slum families of Madras and 60,000 rural landless families in 150 villages in the Tamil Nadu countryside. It organises mass inter-caste weddings to break down discriminatory prejudices. It organises large demonstrations on the problems of its members, and a measure of the political heat it can apply is that Rajiv Gandhi came twice in 1988 to meet different groups of WWF members and listen to their grievances. WWF also organises: child labour rehabilitation centres for working children who would otherwise get no education; training in such skills as needlecraft; advice and counselling on family planning and health care, with an emphasis on preventing infant and mother mortality. This is all part of WWF's objective of increasing consciousness and raising the awareness of its members.

1988 was WWF's tenth anniversary, to mark which it produced a report *Decade of the Forum* (WWF 1988), from which the following quotations are taken.

A decade of Working Women's Forum's success in three southern states can be analysed in development terms, as growth with equity and distribution, betterment in quality of life, increase in women workers' visibility and productivity (through process of credit, employment supportive services) and finally resulting in increased women's consciousness and workers' solidarity. In growth terms, the Forum has increased income/savings, provided cash capital to women workers, credit repayment rate of workers has been over 90%. Implementation of planned health services to the most neglected group of women workers, leads to an identification of a low cost/effective delivery mechanism to improve workers' health and provide women with ability to control reproduction.

Benefits have been distributed to a large population of poor women and access to information/resources have been provided to non-elites. The issues of distributing with equity has been dealt with by conscientising women to confront local power hierarchies as well as exploitative elements such as the money lender, middlemen, exporters and employers. Extraordinary roles of leadership played by women had been the key element. Expanding from urban to rural, from individual workers to piece-rate labour, and from a small effort to mobilization of women, to a trade union for collective action, the Forum has successfully replicated its original urban model (with modification) in 9 different areas. Replication was also possible in different language/cultural contexts.

Being a pioneering effort at the grassroots, its unique organizational structure, leadership roles, mobilization approach, class component of its members and its sheer size, make Working Women's Forum today more a movement . . .

A motivation behind the membership in WWF is a class-centred ideology of a common background and common oppressions – such an ideology transcends barriers of caste, language, culture. The Forum therefore is not strictly an NGO or a union, but a movement marked by the voluntarism of its membership motivated thro' specific class objectives. The Forum stands for a political process of empowerment of poor women, and this empowerment in itself is greater substance to keep its members together, than credit, health or any other material programme. Hence the ideology explains the tremendous growth and continuity of the Forum.

Forum understands its role to be a pressure group, relates itself to Government and other power structures. Two very different functions combined in the activities of the Forum are the flexibility of an NGO and the political character of the trade union. While the NGO character gives it the material support, the trade union character gives it political ideology to sustain its membership and fight for equal opportunities.

The Forum has collected significant experience effectively in re-structuring the living conditions of poor women which could be used by other organizations, particularly in the Asian and African continents. This could take place if WWF opts to train resource persons to start similar movements in other areas.

(WWF 1988, pp.1, 12, 13)

Grameen Bank (Bangladesh)

In all the initiatives so far described an important element has been that of credit. The final example of people-centred development in the South given in this chapter is an organisation that exists to make loans, a 'development bank'.

The Grameen Bank was started in 1976 by a 37-year-old economics professor at Chittagong University, Muhammad Yunus, who wanted to prove, contrary to conventional banking wisdom, that the poor were an eminently bankable social group and that consequently the almost universal demand of banks for collateral, which the poor couldn't provide, was both unjust and bad banking. Initially, Yunus' Grameen Bank Project operated in a village near the university campus as part of his Economics Department's Rural Economics Programme, while the loans were provided and administered by the local branch of the Janata Bank (one of Bangladesh's nationalised banks). The objects of Yunus' Project were:

1. To extend banking facilities to the poor;
2. To eliminate the exploitation of the money-lenders;
3. To create opportunities for self-employment for the vast un- and under-utilised personpower resources;
4. To bring the poor into an organisation they could understand and operate, and in which they could find mutual support and socio-political and economic strength.

GBP was successful and by 1980 seven nationalised banks were involved in running twenty-five branches of GBP in about 300 villages. By the end of 1983 the number of branches stood at eighty-six in 1,249 villages with 58,000 borrowers and a cumulative total loaned of Tk195 million (approximately US$5½ million). In October 1983 the GBP was formally inaugurated as the Grameen Bank (GB). Its subsequent growth has been enormous. By September 1990 there were 754 banks in over 18,500 villages (Bangladesh has 68,000 villages and a population of 110 million, over half of whom are landless) in more than half of the country's districts, with 800,000 borrowers taking loans of US$5.6 million each month for 400 different economic activities. Grameen also gives housing loans, paid back in 10–15 years in weekly instalments. Up to September 1990, 85,500 houses had been constructed with these loans, and US$22.7 million lent. Continuing rapid expansion of GB is planned. Funds for expanding GB to its present size have come from the Bangladesh Bank (in association with the government), IFAD, SIDA, NORAD and the Ford Foundation. Yunus stresses that the bank is self-sustaining at any given size, due to its 98 per cent repayment rate.

The keys to the Grameen Bank's success are:

1. group organisation and group responsibility for the performance of the loans;
2. meticulous staff training – GB currently employs 12,000 people, training 2,000 more in 50–75 batches each year;
3. strict financial discipline using conventional banking procedures;
4. a total commitment to and faith in the villages they are serving.

Grameen specifically targets the landless poor and especially the women among these: about 90 per cent of its borrowers are female. Some villages, with all-women Grameen groups, do not accept male members of the Bank, a locally autonomous decision rooted in the extremely oppressive power structure as far as poor women are concerned. Before loan-giving comes an intensive course of training, group discussions and conscientisation. Groups are encouraged to accept a sixteen-point programme for social justice, group solidarity and women's emancipation. Very importantly, the bank is 75 per cent owned by its borrowers and 25 per cent by the government (though Yunus is seeking to transfer all but 5 per cent to the borrowers also). Only borrowers may own shares; no-one may own more than one share; female shares may not be sold to men.

The benefits brought by Grameen to its borrowers are immense in both a material and non-material sense. The villagers and their children don't starve any more (many did so frequently before); their houses keep out the monsoon; the women have more than one sari and some undergarments; some have started schools with their savings, etc. In addition to credit, GB provides seeds, saplings and oral rehydration kits (for infant diarrhoea) at cost. Yunus now wants GB to branch out into business management and has set up the Grameen Trust. He has his eye especially on the 'corpses' of old development projects which litter Bangladesh – one district has 1,055 broken-down tubewells alone. Grameen has already taken over a large defunct fisheries project originally funded by the UK Overseas Development Administration and rendered it productive in three months. Like the Bank itself, these projects will be jointly owned by the borrowers.

1989 saw the start of Grameen's four-year expansion plan, which is to cost US$125 million. Upon completion the Bank will have a network of 1,000 branches nationwide serving above one million landless poor out of whom 80 per cent are expected to be women. Notwithstanding its ambition, it is a target which on present trends should be amply achieved.

South Shore Bank (USA)

Muhammad Yunus is not a man to keep his Grameen Bank experience to himself, and he is actively promoting similar initiatives in several parts of the world including Burkina Faso, Guinea, Mali, Malaysia (Grameen Trust 1989) and the US. In the US he has acted as a consultant on an adaption of his group lending programme, which is in turn one aspect of a 'development bank' (i.e. one that works for the economic betterment of poor people and communities) designed and established in Arkansas by another remarkable banking institution, the South Shore Bank of Chicago.

The Shorebank Corporation represents a conceptual break-through in the fight against poverty. (It) is among the first development vehicles ever created in this country . . . truly appropriate to the needs . . . in poor communities . . . Shorebank does not exist to make a profit; (it) exists to solve social problems . . . (It) uses the methodology of the private sector to achieve public goals. . . . It is the creative tension between (its) social goals and the bottom line that makes (it) so effective.

(Osborne 1988, quoted in Shapiro 1989, p.1)

Shorebank Corporation . . . was incorporated as a bank holding company in 1972 to foster permanent renewal in blighted neighborhoods. . . . In pursuit of this development agenda it must generate reasonable returns to shareholders and lasting benefits to residents of the communities in which it operates.

(Shapiro 1989, p.2. Joan Shapiro is Senior Vice-President of South Shore Bank)

In 1973 Shorebank took over the South Shore Bank of Chicago, a normal commercial bank in the South Shore area of Chicago, which

in less than a decade had changed from being a diverse, white, middle-class and professional community to an essentially all-black, low to moderate income neighborhood with all the indicators of irrevocable decline into a slum.

(ibid. p.5)

The founder of Shorebank Corporation was a banker, Ronald Grzywinski, who had pioneered the concept of the neighbourhood development corporation based on a bank:

Of all the possible initiators of neighborhood renewal, none better combines financial and managerial resources than a regulated, diversified, deposit-taking financial institution. The depository, a commercial bank or savings association, is known, trusted, legitimate, well-capitalized and self-sustaining. If located within the community, it possesses unusual capacity to be continuously knowledgeable about the neighborhood economy. Most significantly, it converts ordinary bank deposits into development credit. The availability of credit, combined with restored self-confidence, can precipitate the release of local energies, inducing residents to risk their own savings and become personal stakeholders in the future of the community. Through its non-bank development affiliates, the regulated depository can invest equity capital in businesses owned by others, rehabilitate residential and commercial real estate, operate social development programs, attract other private and public investors and, generally, link residents, local government and financial resources into a coherent renewal effort.

(quoted in Shapiro 1989, p.8)

Having defined the concept, Grzywinski and his colleagues set out to test it, raising US$3.2 million in equity and debt and incorporating Shorebank in 1972:

> In its early years Shorebank had to reverse a dramatic outflow of deposits from the Bank and confidence from the community. It did that by simultaneously making the Bank's services competitive, organizing community meetings, extending credit, inducing participation by private mortgage insurers, inviting local residents onto its board, remodeling its facilities, attracting public subsidies to development, retraining its staff ... attracting a community-sensitive housing rehabilitation company, raising investment capital for community organizations and generally heralding the revival of South Shore. . . . (As) an intelligent but inexperienced staff ... (was) trained in banking and development ... it began to make increasingly difficult loans for multi-family rental housing rehabilitation, small business start-ups and commercial district renewal.
>
> (ibid. p.10)

> In 1978 Shorebank capitalized its non-bank subsidiaries ... to diversify the renewal process and engage in activities prohibited to regulated commercial banks in the United States – real estate development, venture capital, non-profit housing and jobs programs.
>
> (ibid. p.10)

Shorebank's success has been remarkable. By the end of 1989 it had total assets of US$184.8 million and a capital of US$11.4 million. The South Shore Bank has extended over US$130 million of market-rate, unsubsidised credit to over 7,000 local business people and residents, weil over half of whom had never borrowed from a bank before with a repayment rate of over 98 per cent. Its real estate development subsidiary had rehabilitated 1,142 units of multi-family rental housing, and had broken ground on a US$10 million shopping plaza. Its non-profit subsidiary has trained 2,218 people and found jobs for 2,734, disbursed US$1 million in low-interest energy conservation loans, and rehabilitated 376 units of housing for the lowest income residents. Housing units completed by all Shorebank companies – banks, real estate and no profit companies – totalled 8,898. Cumulative Shorebank development investment since 1974 had reached US$212 million by the end of 1989.

Shorebank firmly believes that its experience is replicable and has established an advisory subsidiary to deal with the increasing flow of enquiries as to how this might be done. It has itself expanded out of South Shore into four other Chicago neighbourhoods, has established a new bank holding company to promote rural economic development in Arkansas and in October 1990 began collaborating with the Polish American Enterprise Fund to help create a private banking system in Poland. Shapiro summarises its distinctive approach which has proved so rewarding:

> Re-establishing a viable market economy is the most appropriate goal to ensure a neighborhood's long-term revival; bank holding companies, organized for profit as community development corporations, represent an as yet underutilized opportunity for achieving that goal; credit can be organized and used as a poverty alleviation strategy; bank deposits can provide a predictable source of development capital; neighborhood renewal requires 'patient' capital and is inconsistent with short-term profit maximization; at the same time, it is possible to generate financial and social return – profitability and development – over the long-term and with low and moderate income residents as the primary beneficiaries.
>
> (ibid. p.1)

THE PROGRESSIVE MARKET

The South Shore Bank is one of the outstanding examples of a new phenomenon which is becoming increasingly evident in all the mixed industrial economies, one which can best be described as a 'progressive market', whereby economic activities are undertaken through the market for specific ethical, social or environmental purposes as well as for financial return or in pursuit of self-interest.

This market, like any other, comprises (progressive) investors, producers and consumers, the distinctive activities of all of whom are now clearly visible and are becoming significant in comparison to the mainstream market place.

The progressive investor

In the US, the Social Investment Forum, the association for progressive investment institutions, estimates that some US$500 billion is now screened for investment that takes account of ethical or social criteria. The South Shore Bank itself had US$76 million in development deposits in 1989, money deposited with the bank by investors explicitly wishing to support the Bank's community renewal ethos as well as reap a financial return. The same phenomenon is apparent in European countries, where new and significant initiatives like Germany's Ökobank, continue to be launched. In the UK about twenty ethical institutions with different ethical concerns now offer practically the whole range of personal investment services: unit trusts, personal pensions, a building society, other investment plans. Several ethical share issues appealing explicitly to altruism rather than the profit motive, have been oversubscribed, including those of the fair trading corporation Traidcraft and Industrial Common Ownership Finance, which finances worker co-operatives.

The progressive business

On the producer side, industrial co-operatives have always sought to combine social and financial objectives. But in traditional business theory firms are supposed to be run for the exclusive benefit of their shareholders, although it has for some years been recognised that managers are inclined to pursue their own objectives as well.

However, there is another theory of the firm founded on the concept of stakeholders. This recognises that a firm's activity affects many parties: shareholders, customers, employees, suppliers, the local community and the natural environment. The firm therefore has an obligation to each of these parties which should be explicitly reflected in its policy.

The stakeholder theory of business is enjoying something of a revival. Preston and Sapienza (1989) convey the changing climate in a recent paper:

128

The proposition that business corporations can and should serve the interests of multiple *stakeholders* rather than simply those of *shareholders* is enjoying a considerable current vogue. . . . A 1988 Business Roundtable study of 100 major companies found 'widespread recognition that corporate obligations extend to a variety of constituencies or stakeholders . . . that are considered in planning and evaluating corporate policy and action.'

(p.1)

After some econometric analysis of data obtained from the regular *Fortune* magazine surveys, the authors conclude:

Our analysis clearly reveals that multiple objectives – including both economic and social considerations, can be and in fact *are* simultaneously and successfully pursued within large and complex organizations which collectively account for a major part of all economic activity within our society.

(p.19)

It would be idle to pretend that concern for the welfare of all its stakeholders was high on the agenda of more than a handful of firms, but that handful, in rhetoric at least, includes such corporate giants as NCR, Johnson Wax and IBM (which is not to say that rhetoric is always turned into reality when business pressures intensify). Nor are management techniques for handling the complexity of these issues in a competitive environment well developed and there would seem to be a clear role for governments to incorporate stakeholder concepts in company law to force directors to address these issues and to provide for damages in clear cases of neglect.

There are several straws in the wind which indicate how this area could develop: the growth in worker co-operatives, community businesses and other 'local employment initiatives' in many European countries; organisations like Business in the Community in the UK, through which many of the UK's largest companies pledge support for disadvantaged communities; encouragement of employee stock ownership programmes; the growing awareness among many businesses and their management consultants that their most precious resource is their workforce, with all that implies in terms of employee welfare and development.

While these practical expressions of the stakeholder concept are still marginal to mainstream business operation, they have become

129

significantly less so in the last decade. It is not too fanciful to imagine that the continued intensification of this trend could trigger a transformation of business culture, whereby profitability began to be seen as a signal of continuing healthy viability and stakeholder service, rather than the overriding object of maximisation.

The progressive consumer

The most cursory explanation of human nature indicates that most people have strong opinions and preferences about ethical, social and environmental matters. Yet it has for long been routinely and inexplicably supposed that such preferences are irrelevant to people's purchasing decisions. The supposition was self-fulfilling because it resulted in consumers being given no information about these aspects of the products they were about to buy, so that they were unable to take them into account even if they wished to. Consumers therefore unwittingly endorsed the destruction of tropical rainforests and the ozone layer through their purchase of hardwoods, hamburgers and aerosols; contributed to the blinding of young women in Mexican semiconductor factories through their purchase of personal computers; and otherwise were party to a host of other exploitations of people and planet through being unaware of the wider implications of their consumption. In many countries consumers have begun, on a significant scale, to demand information that will remedy this situation.

In itself the demand for consumer information is, of course, not new and several national consumer associations have been purveying such information for many years. A significant step towards consumer solidarity was taken in 1960 when five of the best established associations set up the International Organization of Consumers' Unions (IOCU) to carry out product testing and bring cohesion to the consumer movement internationally. By 1990 IOCU's members numbered 170 groups in sixty-five countries. It had also been instrumental in setting up about twenty networks beyond its formal membership, such as the Pesticides Action Network and Health Action International, which work and campaign in areas such as pesticides, pharmaceuticals, baby foods, food irradiation, biotechnology, toxic waste and tobacco.

Another manifestation of altruistic consumer concern were boycotts of companies perceived as immoral, whether it was Barclays for banking in South Africa or Nestlé for promoting infant formula in Third World countries against World Health Organization guidelines. Some of these boycotts carried clout, especially when governments could be persuaded to come aboard. Banking and trade sanctions in protest against apartheid are estimated to have cost South Africa US$17.5 billion (*Economist* 1990).

Progressive consumerism does not have to be exclusively oriented towards protest and there would appear to be a growing desire among consumers to use their purchasing power positively in pursuit of desired goals. An example of this in the UK was *The Green Consumer Guide* (Elkington and Hailes 1988) to products of comparatively low environmental impact, which shot to the top of the best-selling non-fiction list soon after its publication. It is also exemplified by the two organisations whose profiles follow.

Seikatsu Club Consumers' Cooperative (Japan)

The Seikatsu Club traces its foundation back to 1965 when a single Tokyo housewife organised 200 women to buy 300 bottles of milk to reduce their price.

> What started as a strategy to save money, however, gradually developed over the next twenty years into a philosophy encompassing 'the whole of life'. In addition to cost-effective collective purchase, the club is committed to a host of social concerns, including the environment, the empowerment of women and workers' conditions.
>
> (Seikatsu Club 1988, p.6)

The growth of the Seikatsu Club (SC) has been exponential. By early 1989 it had a membership of 170,000 households, organised in about 27,000 *hans* (the basic local group unit averaging eight members) in eleven prefectures in Japan. 'In the Seikatsu Club we are seeking to empower each and every member with a voice and role in participatory politics' (ibid. p.13). The *hans* elect a representative to their Branch (50–100 *hans*), which develops its own agenda activities and in turn sends representatives to the General Assembly to make policy and elect SC's Board of Directors (80 per cent of whom are women).

> SC is calling on the public to create a self-managed lifestyle in order to change the present wasteful lifestyle, which is a fallout of the present capitalist-controlled society. We believe

131

that the way to improve the quality of life is to create a simple but meaningful existence, and refusing to fall under the having-it-all illusion created by commercial products. To control and manage your own life is a significant factor in realizing a higher quality of life. . . . The objective of SC is to learn how to self-govern society through the self-management of our lives. Our visions for rebuilding local societies derive from this principle. One of our directions is to create local based economies.

(Tokyo Seikatsusha Network 1988, p.3)

In pursuit of this objective, SC has become a formidable commercial enterprise, with a turnover in 1987 of 41 billion yen, (approximately US$300 million) employing 700 people, and member investment of 7½ billion yen (each member invests 1,000 yen per month over the first four years of membership, a stake which is non-interest bearing but returnable on a lapse in membership). SC's principal business centres on the distribution to *hans* of 400 different products, of which sixty are original brands and 60 per cent by value are primary products, like rice, milk, chicken, eggs, fish and vegetables. It is carried out through a unique computer-operated advance ordering system to enable producers to plan in advance and guarantee product freshness.

We refuse to handle products if they are detrimental to the health of our members or the environment. Synthetic detergents, artificial seasoning and clothing made with fluorescents are all off limits, even if members make demands for them. But our commitment to the environment is far more extensive than that. For one thing, the club gets safer produce by cooperating with local farmers. . . . We stand by the belief that housewives can begin to create a society that is harmonious with nature by 'taking action from the home'. And through our purchases and consumption, we are attempting to change the way that Japanese agriculture and fisheries are operated. As a symbolic gesture of societal responsibility for past crimes due to careless industrialization, we buy summer oranges from families with Minamata disease. When the club cannot find products which meet our quality, ecological or social standards, we will consider starting our own enterprise.

(Seikatsu Club 1988, p.8)

This is precisely what they did in the case of milk and natural soap – the Club owns two organic dairies and manufactures several varieties of soap, including one from recycled cooking oil. SC also places great emphasis on direct producer/consumer links to moderate and humanise the market, especially in the area of food production, where consumers regularly visit farmers to inform themselves about or help them in their work.

With the growth in female labour-force participation in Japan, the voluntaristic, housewife-based *han* distribution system came under some stress, as a response to which SC set up women workers' collectives to undertake both distribution and other service enterprises, including recycling, health, education, food preparation, child care, etc. By December 1987, fifty-seven such collectives were giving employment to 1,550 SC members. SC has also established its own not-for-profit insurance company for members.

In their campaign against synthetic detergents, SC members realised the importance of the political process and formed independent networks in different prefectures to contest local elections. In 1979 the first SC member was elected to Tokyo City government and there are now thirty-three councillors in councils in Chiba, Tokyo and Yokohama, all of whom are women. The manifestos of these networks, of which there are now twenty-two, bear a great resemblance to European Green manifestos: very environmental, peace-oriented and anti-nuclear, an emphasis on local participatory democracy, economic and political, and equal status for women. One Kanagawa Network slogan proclaims 'Woman Democracy: Peace, Life, Future, Nature, Earth'.

Another important SC-supported campaign is the Radiation Disaster Alert Network, started after Chernobyl and launched on Hiroshima Day 1986. A community co-operative school and high-tech workers' co-operative invented and manufactured a special geiger counter of which, by March 1987, there were eighty-one in use in twenty-six Japanese prefectures, comprising a nationwide radiation detection network of great sophistication, campaigning at the local and national levels.

SC is ambitious. In its campaign 'From Collective Buying to All of Life' it is planning to contact every household in Japan and recruit 10–30 per cent of them. The political networks are planning at least to double their representation. They have also spread their ideas to South Korea. 'Through such a cooperative community based upon the ideals of the SC in various branches of life, welfare, health, education, culture, environment etc, the present-day rural and urban societies can be regenerated and humanised' (personal communication from Takashi Iwami, an adviser to SC, 15 March 1989).

Council on Economic Priorities (CEP) (USA)

CEP was founded in 1969 by Alice Tepper Marlin, a securities analyst, who was asked to compile a 'peace portfolio' of corporations with the least involvement in supplying the war in Vietnam. Its subsequent research has focused on three areas: peace economy; energy and environment; and corporate responsibility. CEP publishes major studies, monthly research reports and other research.

In the last few years the CEP activity with the highest public profile has undoubtedly been in the area of corporate social responsibility. In 1986 it published a path-breaking work *Rating America's Corporate Conscience* (Lydenburg *et al.* 1986) which rated 130 major US corporations across eight social and ethical criteria and included additional detailed corporate profiles. The criteria were: charitable donations, political donations, military and nuclear weapons-related contracts, position of women and ethnic minorities in the company, social disclosure and activities in South Africa.

In 1988 CEP condensed this information and published its *Shopping for a Better World*, a pocket-sized supermarket guide to 138 companies and 1,300 brand-name products commonly found in US supermarkets across twelve criteria (those above plus environment, animal testing, nuclear power and community). 300,000 copies of this guide were sold in the first nine months, well beyond CEP's expectations. By December 1990, 800,000 copies had been sold. The 1991 edition covers nearly 200 companies and their 2,000 brands. CEP conducted a survey of purchasers, in which 80 per cent of respondents claimed to have changed their shopping habits as a result of information in the guide.

Following the success of its shopping guide, CEP is now preparing to launch an ambitious new Environmental Data Clearinghouse 'which would incorporate data from the numerous local and national sources of environmental information and establish a comprehensive model for reporting corporate social responsibility in this vital area of the environment' (CEP 1990). CEP are now acting as consultants to a new UK group called New Consumer, which is doing a similar analysis of the UK corporate scene, and sharing information with Asahi Shimbun, which has undertaken social responsibility analysis of Japanese companies. New Consumer's major research book, *Changing Corporate Values* (Adams *et al.* 1991) was published in April 1991, with its shopping guide following in September.

Alternative trading organisations (ATOs)

Another manifestation of increasing consumer concern about where, by whom and with what impacts products are made is the growing ATOs movement. These organisations were set up as a reaction against perceived inequities in North-South trade, seeking particularly to give Southern producers a better deal in their transactions with the North by working directly with them and by linking them to sympathetic Northern markets.

'Since their humble and marginal beginnings in the 1960s, ATOs have come a long way. Many Northern ATOs now have multimillion dollar annual turnovers, handle a wide range of products and are reaching many hundreds of producing organisations worldwide' (Tiffen 1990 p. 13). From Finland to New Zealand, there are over forty organisations that call themselves ATOs. 'In addition there are many individuals, groups, and cooperatives running independent shops, with cooperative networks and church organisations providing outlets for products' (ibid.).

ATOs differ markedly from normal commercial trading companies. While they clearly need to remain financially viable, their principal concern is not with their profitability but their ability to give their Southern producers a fair return and security of production. To achieve this ATOs need to work at both ends of the product chain. They give practical advice to producers on such matters as design, appropriate technological innovations and marketing, to help them cater for Northern tastes and markets; and they cultivate those markets, seeking to build long-term commitment among Northern consumers to fair trade, not least by helping them to understand who produces the goods being bought, how and under what conditions, and the difficulties and injustices for the South of the global trading system generally. They act therefore, as mediating organisations, injecting human sympathy and understanding into the otherwise impersonal world of trading relations.

Because ATOs are committed to working with smaller producers, often village-level co-operatives or other groups, the scope for economies of scale in production is obviously limited. But the possibilities for co-operation between them in terms of research, marketing, distribution and other intermediary functions are immense and should help them eliminate some of the waste inherent in market competition. An important step in this direction was the recent creation of the International Federation for Alternative Trade, aiming 'to increase the flow of information, speak out against specific injustices, improve market access and facilitate international partnership' (Tiffen 1990, p.15). Thus as the conventional market becomes in effect a globally integrated operation, so the progressive market is consolidating its international links. It needs to continue to do so with all vigour if it is to be perceived as a real alternative to a world trading system powered by unprincipled market forces.

Promoting the progressive market

The moves towards a progressive market are of great potential importance, because they are based on a new idea which could act as a very powerful force for social change. The traditional view has been that the market-place is the area in which people pursue their narrow self-interests – income, growth, profits, consumer benefits. In order to express their ethical and social priorities they were expected to support charities, vote or be active in political parties, or become involved in pressure groups or other organisations. The distinction between these two realms of activity was more or less absolute: business was hard-nosed and ruthless; consumerism was self-centred and materialistic. On the other side, many social activists were hostile to the market and all it stood for. The progressive market concept directly challenges this dualism, asserting that the great power of the market to express individual preferences can in principle be harnessed to social, ethical or environmental preferences as much as to selfish, materialistic ones.

It is too early as yet to say definitively whether progressive consumers, producers, and investors will fulfil in practice their evident potential for positive social and economic change. If they are to do so, it will be due to the twin action of determined individuals and enabling legislation. The individuals will need to provide the will and the independent research and information organisations to enable them to express their wider preferences with confidence. Governments will need to ensure the appropriate provision of consumer information, the incorporation of stakeholder concepts and concessions for co-operatives into company law, and the right of institutional investors to pursue non-financial investment objectives where their contributing individuals so desire.

Governments in fact are beginning to move on such issues. On the environmental front, for example, Germany, Canada and Japan already have government-backed ecolabelling schemes, with other European countries poised to follow suit. There is no contradiction implied in this combination of individual market preferences and governmental action. All markets are defined and policed by collective consent. What will be removed is the schizophrenic anomaly whereby people in the market-place are presumed to be egotistical and selfish actors, while as citizens in the family, community or polling booth they are widely acknowledged to have powerful moral and social commitments. The progressive market concept relocates the market-place in its human, social context and, in so doing, imbues it with a powerful potential for the transformation of traditional market institutions.

CONCLUSION

It should be clear that all the initiatives which have been explored in this chapter fall squarely within the framework of Another Development outlined in Chapter 5. As one would expect, that framework has been greatly developed in the years since 1975, recently by such works as *The Living Economy: a New Economics in the Making* (Ekins 1986), *Towards a Theory of Rural Development* (De-Silva *et al.* 1988), *For the Common Good* (Daly and Cobb 1990), *Future Wealth* (Robertson 1989), 'Human-scale Development: an option for the future' (Max-Neef *et al.* 1989) and *Real-Life Economics* (Ekins 1992). A detailed survey of these works is beyond the scope of this book but they are all major contributions to a new way of thinking about and acting within an economy to achieve a development which involves 'a purposeful growth of human personality through the release and application of man's creative energies within a collective framework' (De-Silva *et al.* 1988, p.159). 'The conditions for social and economic progress are simply those which release the energies and creativity of the people and transform this creativity and motivation for work into the means of production' (ibid. p.18).

Max-Neef *et al.* (1989) express a similar idea thus:

Human-Scale Development is focused and based on the satisfaction of fundamental human needs, on the generation of growing levels of self-reliance and on the construction of organic articulations (i.e. coherent and consistent relations of balanced interdependence between) of people with nature and technology, global processes with local activity, the personal with the social, planning with autonomy, and civil society with the state.

(p.12)

Fifteen years ago these ideas and the initiatives which flow from them would have been put on the very margins of development theory. Now they are practically universally acknowledged to play an important role. For many they are already perceived as the best, if not the only hope, not only for human betterment but also for achieving the environmental regeneration on which such betterment is increasingly being seen to depend.

7

ENVIRONMENTAL REGENERATION

The earlier analysis of the environmental crisis revealed two principal causes of the destruction of the environment:

1. Wasteful, polluting, resource-intensive consumption patterns of the rich;
2. Environmentally destructive practices of the very poor.

Between these two causes there is a third that partakes of both: a 'development' pattern that takes natural resources away from the poor, who were using them sustainably, and gives them to the relatively rich, who exploit them unsustainably. Where poor people are subject to displacement onto marginal or unsuitable forest land, whether by cash-cropping (e.g. plantations, ranches) or government policy (e.g. Indonesian transmigration programme), their struggle for survival will obviously impact negatively on the environment. Moreover, the use by subsistence-based people of forest resources for their survival is often in conflict with these resources' use for industrial or cash-crop purposes, as countless conflicts from Brazil (e.g. the assassination of Chico Mendes) to Sarawak (e.g. the logging blockades by the Penan people), India (e.g. social forestry in Karnataka, see EDF (1987), and the Narmada Valley Project, already described in some detail) and Thailand, demonstrate.

The cash crops versus food crops issue is described thus in Timberlake (1985):

The widespread planting of cash crops can also cause desertification. First, in situations where croppers are borrowing temperate agricultural practices for large cash-crop monocultures – without making due allowance for the realities of Africa's soil and climate – then the schemes themselves can overburden the land. Second, planting the best land in cash crops, which almost invariably use less labour than food crops, can push large numbers of subsistence farmers and herders onto more marginal land, resulting in desertification.

(p.69)

Norman Myers perceives the small-scale farmer to be responsible for more deforestation than commercial loggers and cattle ranchers combined, while being the least to blame.

> In his main manifestation as the shifted (displaced) cultivator, the small scale farmer is subject to a host of forces – population pressures, pervasive poverty, maldistribution of traditional farmlands, inequitable land tenure systems, inadequate attention to subsistence agriculture, adverse trade and aid patterns, and international debt – that he is little able to comprehend let alone to control. Thus he reflects a failure of development strategies overall, and his problem can be confronted only by a major restructuring of policies on the part of governments and international agencies concerned.
>
> (Myers 1989)

Myers' analysis is essentially confirmed by the World Resources Institute, one of the original framers of the Tropical Forestry Action Plan (see pp. 149ff), who acknowledge that the causes of deforestation include not only population pressure for agricultural land and the demand for firewood and fodder but also:

> skewed land distribution and insecure tenure . . . unsustainable exploitation of forests for industrial timber production and export, and inappropriate government policies regarding land tenure, economic incentives, forest settlement and other population issues. . . . Commercial exploitation is a major cause of deforestation. . . . Large-scale development projects in agriculture and other sectors, including projects funded by international development assistance agencies are major factors as well. As these and other forces reduce the amount of available forest and arable land, poor farmers are forced to move into fragile upland forest areas and marginal lowlands that cannot support large numbers of people practising subsistence agriculture. . . . To hold the poor responsible for this worsening situation is factually and morally wrong.
>
> (WRI 1987 p.10, cited in Colchester and Lohmann 1990 p.6)

Lohmann (1990a) gives a practical example of just this sort of situation from Thailand:

> The proponents of large-scale afforestation schemes in Thailand are using environmental concerns as a smokescreen for the commercialisation of common lands and the destruction of the rural subsistence economy. Hundreds of thousands of local people will be thrown off their lands, many with little option but to encroach on the country's remaining forests thus exacerbating the deforestation crisis. Rural activists are fighting for their livelihoods against multinationals, aid agencies and the Thai business elite who are vigorously promoting the plantations.
>
> (p.9)

Thus a variety of power realities in many developing countries and internationally result in the forcible dispossession or marginalisation of subsistence farmers as more powerful social groups seek to benefit from their resources. Resource degradation is a frequent result. So there is a third significant engine of unsustainability: transfer of resources from the poor, who were using them sustainably, to the relatively rich who do not do so.

The three root causes of environmental destruction can thus be seen not to be very complex conceptually. Similarly their solutions are easily described:

1. *Development and deployment of resource-efficient, less polluting technologies* which enable relatively high standards of living to be sustained at a fraction of current environmental impacts, combined with a willingness on behalf of consumers to forgo consumption that is not environmentally sustainable.
2. *A reorientation of investment* away from resource-exploitation to a massive programme involving the sustainable production of biomass. This is the solution advocated by Agarwal (1985):

> The answer to India's immediate problem of poverty, therefore, lies in increasing the biomass available in nature, and moreover, increasing it in such a manner that access to it is ensured on an equitable basis.
>
> (p.23)

3. *A reversal of policies of dispossession* to those of granting to the poor access to resources, especially land, with incentives for them to be used sustainably.

While these solutions are unproblematic on paper, and there is no doubt that the world community has ample resources to implement them, unfortunately each one is opposed by formidable vested interests which have so far thwarted the sort of response the environmental crisis demands. One example of this is given in Lovins (1989) with regard to the deployment of new energy technologies which could stop the greenhouse effect more or less dead in its tracks:

> The good news is that if we simply pursue the narrowest of economic interests, the energy problem has already been solved by new technologies. . . . The bad news is that most governments and many private sector actors are less committed to market-outcomes in energy policy than to corporate socialism – to bailing out their favourite technologies, many of which are now dying from an incurable attack of market forces. So long as this ideology continues to dominate public policy and the private investments which that policy influences, energy will continue to impose intractable economic, environmental and security constraints on even the type and degree of global development that is vital for basic decency.

> (pp.1–2)

This issue is explored in more detail later in this chapter. Other examples of policies that go against both economic efficiency and sustainability are given in the series of recent publications from the World Resources Institute in Washington DC with titles like *The Forest for the Trees? Government Policies and the Misuse of Forest Resources, Paying the Price: Pesticide Subsidies in Developing Countries* and *Money to Burn: the High Cost of Energy Subsidies.*

The following examples of concrete attempts to address the environmental crisis all illustrate one of the three simple solutions cited. They provide ample evidence of the crisis' essentially institutional and political, rather than technical, nature. They have been grouped under the headings damage limitation; regeneration; and reforming consumption.

DAMAGE LIMITATION

Chipko Movement (India)

The forests of India are a critical resource for the subsistence of rural peoples throughout the country, but especially in hill and mountain areas, both because of their direct provision of food, fuel and fodder and because of their role in stabilising soil and water resources. As these forests have been increasingly felled for commerce and industry, Indian villagers, mainly women, have sought to protect their livelihoods through the Gandhian method of *satyagraha* – non-violent resistance. In the 1970s and 1980s this resistance to the destruction of forests spread throughout India and became organised and known as the Chipko Movement.

The first Chipko action took place spontaneously in Uttar Pradesh in April 1973 and over the next five years spread to many other districts of the Himalayas. The name of the movement comes from a word meaning 'embrace': the villagers, mainly women, hug the trees, saving them by interposing their bodies between them and the contractors' axes. The Chipko protests in Uttar Pradesh achieved a major victory in 1980 with a fifteen-year ban on green felling in the Himalayan forests of that state by order of India's then Prime Minister, Indira Gandhi. Since then the movement has spread to Himachal Pradesh in the North, Karnataka in the South, Rajasthan in the West, Bihar in the East and to the Vindhyas in Central India. In addition to the fifteen-year ban in Uttar Pradesh, the movement has stopped clear felling in the Western Ghats and the Vindhyas and generated pressure for a natural resource policy which is more sensitive to people's needs and ecological requirements.

The Chipko Movement is the result of hundreds of decentralised and locally autonomous initiatives. Its leaders and activists are primarily village women, acting to save their means of subsistence and their communities. Men are involved too, however, and some of these have given wider leadership to the movement. One of these is Sunderlal Bahuguna, a Gandhian activist and philosopher, whose appeal to Mrs Gandhi resulted in the green-felling ban and whose 5,000 kilometre trans-Himalaya footmarch in 1981–3 was crucial in spreading the Chipko message. Bahuguna coined the Chipko slogan: 'Ecology is permanent economy'.

A feature published by the United Nations Environment Programme reported the Chipko movement thus: 'In effect

143

the Chipko people are working a socio-economic revolution by winning control of their forest resources from the hands of a distant bureaucracy which is concerned with selling the forest for making urban-oriented products' (Lamb 1981, p.4).

Sahabat Alam Malaysia (SAM)

Sahabat Alam Malaysia (SAM – the Friends of the Earth organisation in Malaysia) has been caught up since 1986 with the native Penan people of Sarawak in a desperate struggle against logging in the province. In 1983 this logging was proceeding at the rate of seventy-five acres per hour, enabling Sarawak to provide 39 per cent of Malaysia's tropical log exports, which amount to over 50 per cent of the world's total. The logging is systematically destroying the culture and livelihood of the area's native peoples, including the Pelabit, Kayan and Penan peoples.

Until 1990 the SAM Sarawak Office was run by a young Kayan, Harrison Ngau, who had for some years helped the native communities with the problems caused by the logging: pollution, soil erosion, land spoilage and destruction of trees and other forest resources. But when the letters and petitions to government departments which he helped them draft brought no improvement, the Penan people began in 1987 to blockade the logging camps and roads, bringing much of the logging to a halt. In June 1987 SAM Sarawak arranged for a delegation of native leaders to go to Kuala Lumpur for talks with the Malaysian Government. Though these were fruitless the trip and blockades generated considerable publicity at home and abroad. In October 1987 Harrison Ngau and dozens of tribal people were arrested and the blockades broken in a police-crackdown, but they started again in May 1988, especially affecting the operations of the Limbang Trading Company, which is owned by the Minister for Environment and Tourism, Mr James Wong.

SAM itself was founded in 1978 by its present President, S. Mohamed Idris, a local businessman who also started the influential Consumers' Association of Penang (1969), Asia-Pacific Peoples Environment Network (APPEN – 1983) and Third World Network (1984) news agency. SAM's other concerns include resource depletion, loss of indigenous seeds, abuse of pesticides in agriculture and soil contamination. It has an extensive publishing programme: a monthly newsletter in two languages, a bimonthly digest of news bulletins on the environment, a news service and numerous single publications. SAM also pioneered the 'State of the Environment' concept with its *State of the Malaysian Environment* in 1983–4 (SAM 1984).

Environmental Defense Fund (EDF) (USA)

EDF was founded nearly twenty years ago with a campaign that eventually resulted in the use of DDT being banned in the US. It seeks to bring together science and law to prevent and reverse environmental damage. In 1990 it employed ninety staff, over half of whom were attorneys, scientists and economists. It has offices in New York, Washington DC, Oakland (California), Boulder (Colorado), Richmond (Virginia), Austin (Texas) and Raleigh (North Carolina) and had a 1990 budget of US$16 million.

In addition to US-based programmes on water resources, air quality, toxic chemicals, energy and wildlife, EDF has an International Conservation Programme, with attorney Bruce Rich on its staff, often working with a coalition of environmental groups, focusing on the havoc wreaked by the lending of the Multilateral Development Banks (MDBs) of which the World Bank is the biggest. In 1985 their exposé of the disaster of the Polonoroeste (Brazil) projects led to the passing of legislation by Congress (on 19 December 1985) drafted by Rich which:

> directs the US Executive Directors of the banks to urge a series of sweeping environmental reforms. These include increasing environmental staffing and changing the banks' operations to regularly include the environment and public health ministries of developing countries, as well as representatives of non-governmental environmental and indigenous peoples' organizations, in the planning and preparation of projects.
>
> (EDF 1985, p.1)

This law followed two other successful campaigning results:

> the World Bank has, for the first time, halted disbursements on a loan for failure to comply with provisions designed to protect the environment and indigenous peoples. The US Executive Director for the Inter-American Development Bank (IDB), also for the first time, has abstained on a loan for environmental reasons, cutting a loan to Brazil for the continuation of the

Polonoroeste project by more than US$14 million. Since the Polonoroeste disbursements were halted by the World Bank in March 1985, the new Brazilian government has taken steps to bring the project into compliance with the loan conditions consistently flouted by the former, military government. These include demarcating various indigenous reserves, and removing illegal invaders from various protected areas.

(EDF undated, p.1)

In 1986 Barber Conable replaced A.W. Clausen as President of the World Bank. In October Rich and a coalition of fifty-one groups, from North and South, submitted a long memorandum on Indonesian Transmigration, which was being funded by the Bank. In December the project was changed, cutting by 60 per cent the number of people being resettled and, in Conable's words, making 'specific suggestions to promote farm models, environmental protection and socio-cultural safeguards as suggested in your recommendations' (letter from Barber Conable to Bruce Rich, EDF, 15 December 1986, p.1).

EDF and the coalition have had further successes with regard to loans to the Brazilian power sector (the US Executive Director voted against approval of a World Bank loan on environmental grounds); and in getting their own alternative proposal of 'extractive reserves' for rubber tappers and nut-gatherers in the Brazil rainforest accepted by two MDBs including the World Bank. Pilot projects are now being set up, although these were put back in 1988 by the assassination of the rubber-tappers' leader, Chico Mendes, by ranchers wanting the forest to be felled for cattle ranching.

Rich and EDF have also focused on India, working with local groups on trying to change huge World Bank projects in Singrauli and the Narmada Valley. A meeting with World Bank officials when Sunderlal Bahuguna and S.R. Hiremath came to Washington also resulted in modifications to the Karnataka Social Forestry Project. In 1988 Rich noted on the Singrauli case:

> We have just learned that the Bank has caved in and will send a special mission to Singrauli in October to meet with representatives of the 23,000 people displaced by the Bank projects there to prepare an emergency rehabilitation plan. For us, a significant victory!
>
> (personal communication 10 August 1988)

146

Rich's role in the Singrauli case is shown by a leaked World Bank memo of March 11, 1987.

> Bruce Rich has been the principal organiser of this (Singrauli) campaign. . . . Rich and allied environmental groups have contributed to improvements in some of the Bank's ongoing projects: Polonoroeste, Brazil; transmigration, Indonesia; Narmada, India; livestock development, Botswana – these have all been substantially changed for the better, but only after long and bitter public debate.
>
> (Beckman 1987, pp.1, 2)

Very significant in this note is the acknowledgement by the World Bank official that EDF intervention had actually improved the Bank projects, 'but only after long and bitter public debate'. No better confession could be desired of the agonising nature of institutional change.

May 1987 heard Barber Conable publicly announce sweeping environmental reforms, stating that 'if the World Bank has been part of the problem (environmentally) in the past, it can and will be a strong force in finding solutions in the future' (Conable 1987). The Bank has now committed itself publicly to undertaking many of the environmental reforms that EDF has been promoting over the past years. Rich comments drily: 'Since the Bank is an institution known for the gap between its rhetoric and the reality of its operations, the need for intense scrutiny of its activities is redoubled rather than diminished.' (Rich 1987, p.2).

Rich's concern has proved to be only too well justified. Two years later he testified to two US House of Representatives sub-committees about the Bank's 'completely unsatisfactory environmental record: for every case where there appear to be promising changes, we can identify two or three ecological and social debacles of equal or greater scale where the Bank refuses to act even when apprised of the facts.' (Rich 1989, p.3). A month later another Congressional meeting considered the same issues, and Rich reported on James Scheuer's (Democrat, NY) (Sub-Committee Chairman) disillusion at the evidence:

> 'It must be said that the Bank has not institutionalised Barber Conable's rhetoric and Barber Conable's demonstrated concern, both for the environment and for computing the predictable, inexorable environmental damage that these

[Bank financed] projects will cause', Scheuer stated. Indeed, the Sardar Sarovar project was only one of literally scores of ongoing and proposed World Bank ecological debacles that have come to Congressional and international attention over the past two years, debacles that have occurred despite a tenfold increase in Bank environmental staff and a proliferation of new environmental policies, action plans and task forces. '[The Bank's] written assurances don't amount to a hill of beans; they don't exist for practical purposes', Scheuer charged. 'Where do the pressures come from,' he asked, 'pressing down on the World Bank to degrade its own procedures and to bring its own integrity into question?'

(Rich 1990a, pp.305–6)

Later in this article Rich focuses on two of the areas which had caused Scheuer's anger and dismay and which appear again in Congressional testimony by Rich in July 1990 (Rich 1990b): tropical forests and forced resettlement. In this testimony Rich begins by giving credit to the Bank where he perceives it to be due:

Over the past year the Bank has approved a number of environmentally and socially beneficial projects that show that its environmental reform efforts launched three years ago are, to a certain extent, beginning to produce results. A new and critically important environmental assessment policy has been instituted, that, although it needs considerable strengthening, is an important step forward.

(p.2)

Some examples of these projects are given: a US$26 million credit to promote biodiversity in Madagascar through a National Environmental Protection and Management Project; US$600,000 to Kenya to strengthen its wildlife parks; US$237 million for two water resource improvement projects in Indonesia; a US$100 million Beijing Environment loan; and US$18 million for Poland to help the government monitor hazardous wastes.

But Rich warns:

For every new, environmentally positive project we see, we can still point to both ongoing and new projects, often much larger in scale, that are causing completely avoidable and unnecessary environmental destruction and social disruption. In the critical areas of energy and forestry, which represent the Bank's

148

largest and fastest growing lending sectors, respectively, Bank policies and projects are promoting unconscionable ecological degradation. . . . A number of the forestry loans the Bank has prepared over the past year have been little more than massive infusions of cash to promote the accelerated commercial logging of some of the planet's last remaining intact forests. . . . Logging is now rebaptised in Bank Speak as sustainable forest management.

(Rich 1990b, pp.2–4)

To substantiate these charges Rich cites a US$23 million project for Guinea that would open up for timber production two-thirds of 106,000 hectares of pristine rainforests; an US$80 million loan to Côte d'Ivoire to promote timber production in nearly 500,000 hectares of relatively intact rainforest and which would also involve (but does not address) the resettlement of 200,000 people; and a US$30 million deforestation loan to Cameroon. 'These examples show', said Rich, 'that the Bank's current forestry lending for tropical rainforest regions is not only intellectually bankrupt, it is a threat to global ecological stability and to the world's remaining rainforests' (p.6).

Rich and EDF are by no means unique in these criticisms. The normally non-adversarial World Wide Fund for Nature also criticised in strong terms the Guinea and Côte d'Ivoire projects and the US Executive Director actually abstained from approving the latter. Worse, these projects are part of the Tropical Forestry Action Plan (TFAP), the main response of the UN agencies and the World Bank to tropical deforestation, which has been roundly criticised as a whole by environmentalists:

The TFAP is failing to achieve its stated objectives and seems set to accelerate rather than curb deforestation. It is marginalising the rural poor, especially women and forest-dwellers. The TFAP will increase rural impoverishment.

(Colchester and Lohmann 1990, p.86)

Yet seventy-three countries accounting for 85 per cent of all tropical forests have expressed an interest in participating in TFAP, which envisages the spending of US$8 billion over five years. World Bank spending alone on forestry could rise to US$800 million by 1992/3. Rich summarises the situation: 'The plan is well on track to mobilise billions of dollars for forestry projects in every country in the world with remaining tropical forests. Environmentalists around the world now fear that an ecological Frankenstein has been unleashed' (Rich 1990a pp.309–10).

On forced resettlement Rich estimates that of January 1990 '1.5 million people were being forcibly displaced by over 70 ongoing Bank projects and proposed projects currently under consideration may displace another 1.5 million' (ibid, p.313). Many are in a similar situation to the Narmada oustees: the Kedung Ombo dam in Java displaced 20,000 people, the Ruzizi II Regional Hydroelectric Project forcibly resettled 12,500 in Zaire and Rwanda; Kenya's Kiambere Hydroelectric Project displaced 6,000, and so on. In each case, consultation with and compensation for the dispossessed was either inadequate or totally lacking.

Overall Rich's evidence on the World Bank record, since and despite its rhetorical green awareness, is one of human and ecological tragedy. This is all the more serious in that the Bank's ability to inflict such tragedy was greatly increased in 1988 when its lending capacity was nearly doubled by a US$75 billion capital increase. The four Multilateral Development Banks (MDBs) are now lending more than US$32 billion annually for projects worth well over US$100 billion overall. The destructive capability of such sums of money is enormous.

That MDBs can do better than the World Bank's present performance, and that Rich can recognise and applaud improvement when he sees it, is evidenced by his testimony on the Inter-American Development Bank (IDB) to the same Congressional committees:

> The IDB has distinguished itself among the MDBs as the bank which has made the broadest commitment to incorporating environmental and social considerations fully into loan operations, at all levels of the institution. The IDB deserves special credit for having recognised and acted on the principle that direct consultation with and involvement of local non-governmental organisations (NGOs) in project work is central to the environmental soundness of projects.
>
> (Rich 1990b, p.15)

Rich goes on to compare favourably an IDB project in the Brazilian Amazon with a World Bank one in the same area.

It is clear that the MDBs in general, and the World Bank in particular have a long, long way to go before their lending is for anything approaching 'sustainable development'. That they have started out on that road at all is in no small measure due to Rich's and EDF's lobbying and NGO-coordinating work, in recognition of which the United Nations Environment Programme recently added Rich to its Global 500 list of outstanding contributors to environmental protection.

REGENERATION

Green Belt Movement (Kenya)

The principal promoter of the Green Belt Movement has been Professor Wangari Maathai, who was born in Kenya in 1940 and received her doctorate in 1971. She became head of veterinary anatomy and Associate Professor of Anatomy at Nairobi University in 1976 and 1977 respectively.

Maathai has also long been active in the National Council of Women of Kenya, of which she has been Chair since 1980, and it was in the National Council of Women that the idea of the Green Belt Movement, a broad-based, grassroots tree-planting activity was born. Its first trees were planted on 5 June, World Environment Day, 1977.

The Green Belt Movement grew very fast. By the mid-1980s Maathai estimated that it had about 600 tree-nurseries, involving and earning income for 2,000–3,000 women; had planted about 2,000 green belts of at least 1,000 trees each, involving about half a million schoolchildren; and had assisted some 15,000 farmers to plant private green belts. Maathai is currently taking forward a proposal with the United Nations Environment Programme for a Pan-African Green Belt Movement, to spread the successful Kenyan experience to twelve other African countries.

Through the planting of Green Belts, the Movement seeks to achieve many different objectives including: avoiding desertification; promoting the ideas and creating public awareness of environment and development; providing fuelwood for energy; promoting a variety of trees for human and animal use; encouraging soil conservation and land rehabilitation; creating jobs in the rural areas especially for the handicapped and rural poor; creating self-employment opportunities for young people in agriculture; giving women the positive image appropriate to their leading role in development processes; promoting sound nutrition based on traditional foodstuffs; carrying out research in conjunction with academic institutions; developing a replicable methodology for rural development. In its first ten years all of these objectives were realised by the Movement to some degree.

In 1989 Maathai's position and that of the whole Green Belt Movement in Kenya was threatened by her opposition to a plan to build a world media centre, including the highest projected building in Africa, on Nairobi's principal inner-city park which was much enjoyed by the city's poor and children. The new centre was a joint project between the Government of Kenya and one of Robert Maxwell's companies and was to feature a more than lifesize statue of President Arap Moi. As Maathai mobilised her contacts against the project and international pressure especially mounted, she was vilified and placed under virtual house arrest. When the Norwegian ambassador, whose country had significantly backed the Green Belt Movement, protested he was summarily sent back to Norway and pressure was exerted on the Movement itself. In November 1990 Maathai herself was prevented from returning to Kenya after a trip to the US. The whole story is a classic example of the connection between human rights abuse, 'prestige project' development and unsustainability.

José Lutzenberger

José Lutzenberger was born in 1927 and is a Brazilian agronomist and engineer who worked for fifteen years for the chemical company BASF, but left in 1972 to start a vigorous and successful campaign against the over-use of agrochemicals. In the ten years after 1978 the use of such chemicals in his home state of Rio Grande del Sul had fallen by more than 70 per cent, which largely reflects his tireless work with local farmers and their associations on 'regenerative agriculture' – the process of increasing the fertility of the soil through food production rather than the reverse. His work in this field has made him an acknowledged expert in soil science and organic fertilisers.

Agriculture is only one of the concerns of this man who is widely known in Brazil as the father of its environmental movement. He is also a trenchant and effective critic of the many projects of the World Bank in Brazil which contribute to deforestation and other environmental destruction, and can share the credit for both the withholding by the United States of some loans for such projects and for the new environmental awareness of the Bank. He is equally critical of Brazil's programme of domestic reafforestation with inappropriate species.

As an engineer Lutzenberger has worked with many of the factories in his area to ease their waste-disposal and pollution problems. There is a wide range of industries involved, including tanneries, slaughterhouses, soya bean oil production, mining by-products and sewerage. He also was a consultant to the Riocell cellulose plant which converted its plant from a major polluter to one of the cleanest in the world.

These activities in Brazil were combined with an international speaking schedule that takes him regularly to the US and Europe. In 1987 for example he spoke at the environmentalists' World Bank conference and visited Austria as a guest of the Austrian Academy of Science and other institutions. In 1988 he contributed to the counter-IMF/World Bank meetings in Berlin and gave the Scott Bader Commonwealth lecture in the UK, organised in association with Friends of UNESCO. He also appears regularly on television in Brazil and on his trips abroad.

In all these areas Lutzenberger's life illustrates the sort of systematic application of scientific and technical knowledge to environmental problems, which will need to become widespread if the global ecological crisis is to be effectively addressed. In March 1990 Lutzenberger was made head of the newly created Special Environment Secretariat of the Brazilian Government 'to draw up the country's environmental policy in a move that heralds a break with the chauvinistic and contradictory attitude of the outgoing government' (Rocha 1990).

Asociacion ANAI (Costa Rica)

ANAI was founded in 1973 by Dr Bill McLarney, an American, then 30 years old, with a degree in biology and doctorate in fisheries. 'ANAI's purpose is to further the cause of earth stewardship in the tropical lowland environment of Atlantic Costa Rica – and in so doing develop methods and concepts which may be of use elsewhere in the beleaguered tropical lowlands' (McLarney 1985, p.4).

From an original seven-hectare site, six hours on foot from the nearest road, ANAI's operation has expanded to include the entire canton of Talamanca (3,000 sq km and 25,000 people), one of the poorest regions in Costa Rica. ANAI's first project was agricultural diversification out of the cacao monocrop, using a 110-hectare farm as a base for field trials, half of which consists of virgin tropical forest which has been placed in a private forest reserve.

> The fundamental thrust of the work is to test and make available to the campesinos of Talamanca genetic materials of the very best tropical crop plants, particularly tree crops. Today the ANAI farm constitutes a virtually unique genetic resource with numerous varieties of over 100 species of fruit trees alone, including disease-resistant root stocks, improved varieties of familiar fruits, and little known Amazonian and southeast Asian fruits from climates similar to Talamanca's. In addition to fruits, collections are being made of lumber trees, nitrogen fixers, spices and pharmaceuticals. These materials are routinely made available to the Costa Rican government and private institutions, but the primary beneficiaries are the campesinos of Talamanca.
>
> (Todd 1987, p.2)

This process led to ANAI's agroforestry project, with the establishment of twenty-five community nurseries throughout Talamanca's principal communities, in which farmers not only produce seedlings but get training in all the skills necessary to manage and propagate the wide variety of species. About 6,000 people have benefited directly from this programme, planting out 1½ million trees and a comparable number of annual and biennial crops. To develop their business skills and prospects, ANAI has helped create a regional procurement, processing and marketing association (APPTA) with a community-level internal organisational structure.

Future plans in this project include more nurseries, the establishment of two educational farms, development of processing, marketing and business training skills and the training of a cadre of 'barefoot agronomists', one or two from each community of 50–100 families, concentrating on reforestation, natural forest management, investigation, training and helping get environmental concerns on the people's agenda as they plan and create their future, 'work which is hopefully leading to greater community autonomy, greater awareness and appreciation of the environmental aspects of development and significantly increased capabilities to direct and effect the positive change that they want to achieve'. (personal communication from J. Lynch, 17 October 1990).

The other major ongoing ANAI programme, begun in 1984,

comprises two mutually reinforcing parts: a wildlife refuge, now recognised by presidential decree, covering nearly 10,000 hectares of some of the ecologically most important land in Costa Rica, and a land titling project for some 500 families in a similar-size 'buffer zone' area round it. By getting a title deed of the land for its peasant farmers, it is hoped that the buffer area will be managed more sustainably and that areas of private forest will be conserved. The response from the peasant farmers to this project, none of whom could have afforded the titling process by themselves, was enthusiastic. The surveying is now done and the granting of titles only awaits completion of bureaucratic procedures. Already there are important spin-offs: a planned destructive new road through the refuge was stopped; and the sound land management of a further 3,500 hectares by a neighbouring Indian community, advised by ANAI on economic and cultural methods, linked to conservation, has considerably increased the area under effective protection.

Bill McLarney writes:

> Although not conceived as such at the start, what has evolved in Gandoca/Manzanillo (the refuge) is very reminiscent of the Biosphere Reserve concept – a core of wildlands ringed by areas accommodating various degrees of human activity and management, with human communities integrated into the conservation strategy. An important future goal is formalization of this through elaboration of a management plan.
>
> (McLarney 1988, p.2)

ANAI has successfully implanted into its local area. Four out of the six-person Board of Directors are Costa Rican, and three are Talamancans, as are most of the staff of twenty-three, although it remains true that the two North American Board members, James Lynch and Bob Mack, remain President and Treasurer respectively and still provide the main leadership. Bill McLarney now spends more time in the US, fundraising for ANAI and doing other PR work. ANAI's work was specifically commended in the *Time* 'Planet of the Year' issue (2 January 1989).

Many volunteers have helped ANAI over ten years and over 7,000 people have benefited from it. Chief funding for its US$600,000 budget has come from ACORDE (Asociacion Costarricense de Organizaciones de Desarrollo) and others. Todd concludes:

(ANAI) ultimately implies the transformation of what has become Costa Rica's poorest canton into one of the most stable farm areas in Central America ... with concomitant environmental benefits through the conversion of Talamanca farms to agroforestry systems, with a resulting decrease in pressure for the clearing of natural forests.

(ibid. p.7)

Lynch realistically comments: 'Population pressure, mostly due to immigration, and even worsening prices for the crops that are grown, make the last quote more of an objective than a current reality' (personal communication, 17 October 1990).

REFORMING CONSUMPTION

Chapter 5 has already had some discussion of the concept of 'the progressive market' and, in particular, the emerging power and potential of the 'the progressive consumer'. In no field is this more true than in that of the environmentally aware, or green, consumer.

Almost all industrial countries have seen a significant spread in recent years among consumers of the awareness of the ability of their purchasing power to support environmentally sound products and business practices. The UK has been especially significant since *The Green Consumer Guide* (Elkington and Hailes 1988) was published in September 1988 and sold 300,000 copies within a year. Enormous activity on the producer side followed, with high level oversubscribed conferences on how to attract the green consumer and numerous 'green product' innovations. Detergents, batteries, motor cars, supermarkets – all these industries made significant green pitches during 1989–90. Mainstream business and marketing magazines kept up the steady exhortation to business to respond to this new consumer pressure as a business opportunity. Even the government felt moved to respond to the new mood and the Department of the Environment (DOE) introduced a discussion paper on the theme of ecolabelling (DOE 1989) as well as commissioning a wide ranging report on the economy and the environment (Pearce *et al.* 1989). In Europe ecolabelling was first introduced by West Germany with its Blue Angel Scheme, and it may be that that country is Europe's most environmentally aware nation. If so, it is largely due to the energy and activities of such people as Dr Maximilian Gege.

Maximilian Gege (Germany)

Maximilian Gege was born in 1945 and was the director of management planning and environment for a Hamburg business. In the late 1970s he became increasingly aware of environmental problems which led to activism in citizens' groups for clean air, against roads, etc. In 1983 he invented the concept of 'environmental advisers' and in 1984 set up AUGE (Action Association for Environment, Health and Food) to promote this:

Environmental advisers attempt to realize the following main objectives:

- To awaken in people a greater environmental awareness and motivate them towards a less environmentally harmful consumer behaviour;
- To increase the demand for environmentally sound products with low emissions;
- To further the development of environmentally sound products by talking with manufacturers;
- To implement behavioural changes that help to reduce environmental pollution;
- To make cost-saving suggestions for households (up to 2,500 DM/year) and for communities, which also reduce environmental burdens (ie reduction of solid and liquid wastes, lower costs through reduction of water and energy consumption, etc.);
- To show how a better environmental policy not only helps to solve current environmental problems, but to improve the community image.

(AUGE 1986, p.7)

From the start Gege conceived the environmental adviser as a new professional. The first three were employed in Hamburg in collaboration with a conservation group. In co-operation with AUGE, a Swiss firm was brought in to give training advice and unemployed scientists started to be retrained as environmental advisers. Extensive networking and promotional work – including publication of a book *Öko-Sparbuch für Haushalt und Familie* (Ecology Savings Book for Household and Family) (Gege *et al.* 1986) which sold 50,000 copies, royalties to AUGE – meant that by 1987 nearly eighty environmental advisers were employed in the FRG.

The success of the environmental adviser concept in West Germany meant that Gege won a commission in 1986 from the European Commission to study its feasibility in England, Spain and France as well, where fifty environmental advisers are now working. The concept has also spread to Luxembourg, Austria, Denmark, the Netherlands and Switzerland, with contacts in Hungary, Canada and other countries. By 1991 about 1,000 environmental advisers were working in Germany and about 500 in other countries, only four years after Gege invented the concept. Courses of study for these posts have been set up at universities and other institutions, and Gege has also developed a large-scale course of self-study to qualify as an environmental adviser which already has about 500 students.

In Germany AUGE's activities were further developed with a massive water-saving programme for Hamburg; an environmental computer program that helps households save money; collaboration with a large mail-order firm Otto Versand to develop an environmentally-sound catalogue; and a nationwide competition, in search of the environmentally soundest household for 1988–9.

This competition was a most ambitious and successful affair. Twenty-three million brochures were distributed to German households, with the declared goal of promoting ecological behaviour by every citizen, resulting in the return of 350,000 questionnaires. This makes it probably the largest mass ecological conscientisation initiative ever undertaken. Winning households from different parts of West Germany were presented with DM20,000 cheques, normally by the State Environmental Minister, with great attendant publicity. All the recipients were women – financial recognition at last for good housekeeping.

To promote the environmentally-minded enterprise, Gege and Georg Winter founded in 1985, BAUM (Bundesdeutschen Arbeitskreis fur umweltbewusstes Management e.v. – German Association for Environmentalist Management) of which Gege became Executive Director in 1989. BAUM is an initiative for the business community trying to promote a holistic concept of environmental awareness in corporate management. Gege also worked with Georg Winter on a book in this field, entitled *Das Umweltbewusste Management* (Business and the Environment) (Winter 1988, also translated into French and Spanish). At present, BAUM has nearly 300 members, most of which are small and medium-sized companies, but there are also some large companies operating on a global scale.

In just a few years, BAUM made itself a name in the Federal Republic of Germany as a multi-sector, non-profit and non-party organisation that has played a decisive role in establishing a movement for environmental management. Since then, major steps have been taken both in the private and in the public sector for reorientation of thinking and action.

There is an increasing number of companies which are gaining market success by changing over to environmentally sounder products and reducing their costs by programmes to save energy, water and raw materials. Waste disposal costs and risks are increasingly being reduced by modern waste avoidance strategies and recycling. Staff are showing increased motivation and initiative. The attitude of the general public, formerly very critical of industry, has become more positive.

The activities of BAUM are based on the 'Integrated System of Environmental Management' which has been developed, tested and successfully used in medium-sized companies. The management model is based on the need to fulfil the economic and ecological requirements simultaneously in all functional areas and at all levels of the company. Another book, which Gege co-authored with K. Apitz and published in November 1990, is entitled *What Managers Could Learn from a Greenfly* (Apitz and Gege 1990) and stresses the need for the human economy to take sustainability lessons from nature.

BAUM has a major role in the research project 'Environmental Business Management' of the German Ministry of the Environment and the German Environmental Agency. BAUM has developed environmental guidelines for the Environment Department of one of the German Länder, to be applicable for all public procurement activities of that Land.

All the projects conducted by BAUM so far with private- and public-sector organisations have shown that ecological measures also give economic benefits. They improve the cost-effectiveness of the company, or show the company new products and new markets.

Yet another initiative of Gege's launched in 1990 is the AUGE Children's Environmental Club, a systematic multi-media attempt to raise children's consciousness of environmental issues. It comprises a magazine, videos, environmental holidays, an environmental protection board game, teaching materials and a touring 'Enviro-mobile'. Gege is concerned to improve the quality of what is available to and marketed for children and young people, especially with regard to food, school facilities and free time and is setting up discussions with industry to achieve this. Characteristically ambitious, Gege hopes that by post-1992 the Children's Club will have become a European affair, and is establishing a European network to this end.

From these organisations emerges the 'grand plan' behind Gege's activities: he is seeking systematically to conscientise, inform and persuade people about environment protection at each of the most critical human/environment interfaces: childhood and youth (Children's Club); consumers/households (AUGE); business (BAUM); government (environmental advisers). Gege's business skills seem to have ensured that each of these dimensions has been most professionally approached and his effectiveness has been recognised by several environmental awards. It is an impressive record that could well serve as a model for other countries.

Consuming less energy

Energy consumption is directly responsible for at least three of the critical environmental problems currently facing humanity: the greenhouse effect, half of which is caused by carbon dioxide emissions from burning fossil fuels; acid rain and other air pollution from the sulphur, nitrogen and hydrocarbon emissions of power stations and motor-cars; and radioactive wastes and accidents from nuclear power plants.

The Intergovernmental Panel on Climate Change has recommended a minimum of 60 per cent decrease in carbon dioxide (CO_2) emissions as necessary to stabilise the earth's climate. The US, which in 1987 was emitting 1.7 times as much CO_2 per head as West Germany, 2.4 times as much as Japan, 9 times as much as China, 26.5 times as much as India and 168 times as much as Zaire (all per head) (Brown et al. 1990, p.19) clearly has a major obligation to lead the way in decreasing emissions.

Estimates as to what 60 per cent reductions in CO_2 emissions might cost vary widely. The Council of Economic Advisers to the US President put the figure for the US at US$3.6 trillion 'under pessimistic scenarios of available fuel substitutes and increasing energy efficiency' to achieve a mere 20 per cent reduction by the year 2100 (CEA 1990, p.214). The Rocky Mountain Institute believes that 'the famous economists who issued these estimates got the amount about right but the sign wrong: saving enough fossil fuel to cut CO_2 emissions by 20% would not cost but save the US about US$200 billion a year' (RMI 1990, p.1). (The US$3.6 trillion present value is about the same as a US$200 billion annual expenditure for 60 years at a 5 per cent discount rate).

RMI backs this astonishing statement with nearly ten years of research and consultancy into energy conservation and efficiency which has won as clients numerous electricity companies and others from all over the world.

> Many US utilities are now saving lots of electricity at total costs of about a tenth of their electricity prices; at some utilities saving electricity is now the most profitable part of their business. As for industry, many companies, even those that already halved their fuel intensity since 1973, are saving another half or more with payback times under two years. Prototype cars more than three times as efficient as the present fleet apparently cost about the same to build as today's cars. As better technologies and delivery methods become available, saving energy is getting cheaper almost by the day. . . . Entrepreneurs take note: there's a half-trillion-dollar a year market here.
>
> (RMI 1990, pp.2, 6)

In another paper (Lovins 1989) Amory Lovins, RMI's Research Director, has given many examples of how the market is already taking advantage of these opportunities. Thus:

> since 1979 the US has gotten more than seven times as much new energy from savings as from all net increases in energy supply; of that new supply more has come from renewables than from non-renewables; . . . because of the reductions in energy intensity achieved since 1973, the annual U.S. energy bill has recently been approx. $430 billion instead of approx. $580 – a saving of $150 billion a year.
>
> (ibid., p.2)

However, Lovins says that even these impressive results are a fraction of what the market could deliver if it were not for lack of information among consumers about the possibilities and downright obstruction by vested interests:

Wastefully used, depletable, environmentally damaging and egregiously uncompetitive energy options have received government succour lavishly and continuously. The triumph of narrow private interest expedience over macroeconomic efficiency, especially in the United States, beggars description.

(ibid, pp.2,9)

Lovins could have just as easily cited the UK government for its extraordinary decision to continue to subsidise new developments of nuclear power in the UK after the electricity industry had had to be privatised minus the nuclear component, because London's Stock Exchange decisively rejected nuclear electricity, despite the offer of a generous government subsidy to be derived from a 'nuclear tax' on electricity consumers.

Policy-makers' continuing obsession with nuclear power is a serious threat to effective responses to global warming, because of the danger that its enormous cost will crowd out, as it always has, investment in efficiency, conservation and renewables. As Keepin (1990) has shown: 'At the present time in the United States, each dollar invested in efficiency displaces nearly seven times more carbon than a dollar invested in new nuclear power' (p.304). A recent study on Sweden further underlines the irrelevance of nuclear power to solving global warming: 'In the most nuclear intensive country in the world (on a per capita basis), emissions of carbon dioxide and other substances can be lowered while switching completely away from nuclear power' (Leggett 1990, p.317).

If governments actually took it upon themselves to encourage energy efficiency and conservation rather than getting in its way, so that the market had a favourable wind, energy consumption would tumble. Such encouragement would cost money, of course, at least until the market was primed. The Gulf Crisis has provided an object lesson in how money can always be found when the will is there:

The response to the Gulf crisis indicates how quickly funding and resources can be made available given the political will. In this light, it is particularly telling that the $411 million being spent in the US on all federal energy conservation for the whole of 1990 is equivalent to only ten days of expenditure on the US forces now in the Gulf. Similarly, the British government spends £15 million on energy efficiency annually, while it is costing 1 million a day to keep British forces operating in the Gulf.

(SF 1990, p.14)

As the crisis deepened, the UK Chancellor of the Exchequer apparently 'insisted that whatever the state of the economy or the pressure on public expenditure, funds would be found for British forces in the Gulf' (*Guardian* 2 January 1991, p.12). Now *there* is political will. What revolution in perception is necessary for the collapse of the Earth's life-support systems to be perceived to be as threatening as Saddam Hussein?

CONCLUSION

Of course the initiatives and projects which have been described here represent a tiny fraction of the popular action which is being mobilised around the world by increasingly serious and widespread environmental concern. In its report to the Brundtland Commission, the Environment Liaison Centre in Nairobi reported that

> The NGO response to the growing challenge of sustainable development, though hindered in recent years by a deterioration in the wider institutional context, has been decisive. NGOs have swung into action in increasing numbers and at all levels from the local to the global, taking up a lengthening list of environment-development issues; developing innovative forms of cooperation between organisations and undertaking a broad range of tasks from advocacy to the initiation of pioneering projects.

(ELC 1986, p.4)

These popular initiatives badly need to be combined with firm governmental and intergovernmental action across the whole range of issues with implications for environmental sustainability. The earlier analysis permits the outline of a possible scenario incorporating such actions.

First, the North accepts its essential responsibility for the environmental crisis and institutes a massive programme of conservation, resource-efficiency and pollution control in its own countries, perhaps through the imposition of annually increasing environmental taxes. This would certainly result in major lifestyle changes, such as a fall in the use of the private motor-car.

The North also recognises the constraints on sustainable

development in the South caused by the world economic system, and undertakes its systematic reform involving debt cancellation and fairer trading relationships (involving issues such as commodity prices, protectionism against Southern manufacturers, corporate codes of conduct, exploitation of the global commons). It also agrees to the concessional transfer of clean, efficient technologies for appropriate Southern industrialisation, and of resources for Southern environmental regeneration.

However, this programme of reform has fundamental conditions for Southern elites, expressed by the words justice, democracy and sustainability. Justice demands the return of illegal flight capital from Northern banks to its countries of origin to fulfil the development tasks for which it was intended. It also demands a recognition of peasant and indigenous land rights through comprehensive and effective programmes of land reform. Democracy demands that people become the controllers of their development rather than its passive instruments or, worse, its victims. This involves the option of rejecting certain development patterns as well as full participation in those chosen. Sustainability involves absolute respect for and conservation of critical global resources such as rainforests as well as rigorously sustainable use of all renewable resources and strict adherence to internationally agreed emission quotas.

Such a programme would cost the North large sums of money which could probably only be found by plundering arms budgets. It would also include lifestyle shifts and a drop in Northern incomes. It would cost the Southern elites their autocratic power and many of their Northern lifestyle trappings. It would give new life and hope to the rural poor worldwide. It would also give the human race a secure future. It remains to be seen whether this final benefit can muster the political will to overthrow the vested interests that currently stand between this scenario and reality.

So far, as has been seen, the political will has been notably absent. There can be few greater examples of lack of vision in world 'leaders' than that, despite their access to the very latest scientific evidence, they have trailed far behind their peoples in recognition of the environmental crisis, which is likely to be the most important political and human issue of the 1990s. It is only a few years since President Reagan was blaming pollution on trees and the UK Environment Secretary Nicholas Ridley called the ecological critique 'intellectually bankrupt'. Once again it has been ordinary people working through largely voluntary organisations who have acted decisively for human wellbeing, while the established power structures were either blind to the perils or actively promoting them.

8

FURTHER ASPECTS OF HUMAN DEVELOPMENT

The illustrations of the new model of economic development which were given in Chapter 6 placed considerable emphasis on non-monetary variables in terms of both their objectives and criteria of success. Such objectives could include health, education, housing, appropriate science and technology, and enrichment of both cultural and spiritual life. It is not surprising that this holistic approach to economic development shares many features with projects or initiatives which are more narrowly focused on one of the above sectors. Several of these projects in different fields will now be described in order to indicate further applications and achievements of the model which is here being presented.

HEALTH

The work of the six individuals discussed here is included in order to provide an overview of the causes of some of the most pressing health problems in the world today and offer feasible approaches to their solution. The areas chosen are health in poor countries; essential drugs; an alternative (homeopathy) to the dominant Western medical monopoly (allopathy); a new approach to Western systems of health care; and the validity of Third World scientific research.

The importance of these five areas is so obvious that their choice needs little justification:

- The causes of and best response to Third World health problems is one of the acknowledged critical global challenges. David Werner has contributed as much as anyone to both analysis of cause and development of cure.

- Drugs have an essential role in any health-care system, but in many countries, especially in the Third World, they are a mixed blessing. Useless or dangerous drugs are common and their ill-effects are compounded by widespread ignorance about drugs among consumers, dispensers, and prescribers. Bangladesh was the first country to tackle this problem, largely through the inspiration of Dr Zafrullah Chowdhury and his organisation Gonoshasthaya Kendra, 'the People's Health Centre', which has sought both to promote an 'essential drugs only' policy and to integrate these drugs into a broader programme of health and development.

- The Western medical model is experiencing both a crisis of effectiveness against such diseases as cancer and AIDS, and a crisis of confidence among Western publics yet it continues to be promoted aggressively in Third World countries at the expense of more appropriate traditional health-care systems. George Vithoulkas' revival of the credibility of homeopathy is an important step in the re-establishment of generally accepted plurality in approaches to health care.

- The crisis in the Western health-care model is not only in terms of perceptions of efficacy. It is also institutional and financial. The Black Report (Townsend and Davidson 1982) concluded that despite spending an ever-increasing percentage of their national income on health services, rich industrial countries such as the US, UK, France, Sweden and Germany have been unable to demonstrate satisfactorily that much higher spending is clearly related to much better health. The United States medical system fits the Black Report description very well. The objectives and experience of the Gesundheit Institute go to the heart of the causes of its malaise in an extremely radical way.

- Finally, there is the question of Third World knowledge and research. The problem is not just that traditional and important Third World knowledge is being disregarded, devalued and destroyed, but also that Third World scientific research is frequently discounted, resulting either in its findings being ignored or in requiring them to be revalidated by Western research institutions. This makes it very difficult for Third World institutions to acquire the resources or reputation to develop successfully. The syndrome is well-illustrated by the case of Drs Aklilu Lemma and Legesse Wolde-Yohannes and their work on the endod berry as a preventive against bilharzia.

A NEW WORLD ORDER
David Werner (USA)

One of the most widely read authors in the world today is a disabled writer/artist who devotes his words and pictures to aiding millions of people around the world who are handicapped by physical or mental limitations, disease, poverty or lack of education.

He is David Werner, 53, of Palo Alto, California. His first major book, *Where There Is No Doctor* was published in 1977 and there are in print today a good deal more than two million copies in more than 40 languages, including pidgin English. It's a training manual for health workers in rural areas of poor countries where physicians are not available. In use in 150 countries, it is the world's most widely used primary health-care manual for community health workers. Along with his other major health books, the manual assures Mr Werner a place as one of history's greatest health educators.

. . .

In 1982 there came *Helping Health Workers Learn*, a book of methods, ideas and aids for instructors at the village level. And in 1987 Mr Werner published *Disabled Village Children*, which is a 670-page manual for health workers, rehabilitation workers and the families of children who have impairments.
(Allen A. and G. 1988, pp. 175, 178)

The introduction to the English edition of *Where There Is No Doctor* sets out the book's essential message:

This handbook has been written primarily for those who live far from medical centers, in places where there is no doctor. But even where there are doctors, people can and should take the lead in their own health care. So this book is for everyone who cares. It has been written in the belief that:

1. Health care is not only everyone's right, but everyone's responsibility.
2. Informed self-care should be the main goal of any health program or activity.
3. Ordinary people provided with clear, simple information can prevent and treat most common health problems in

their own homes – earlier, cheaper and often better than can doctors.
4. Medical knowledge should not be the guarded secret of a select few, but should be freely shared by everyone.
5. People with little formal education can be trusted as much as those with a lot. And they are just as smart.
6. Basic health care should not be delivered, but encouraged.

(Werner 1977a)

David Werner is disabled himself having Charcot-Marie-Tooth syndrome, which causes a progressive atrophying of the muscles. His decades-long involvement with health-programmes for villagers began in Mexico while he was on vacation as a high-school biology teacher. His work is rooted in the principle that lay people can and should be responsible for their own health care. Now his international standing is such that he has been a consultant to the Pan American Health Organization, the Peace Corps, UNICEF, the World Health Organization (WHO) and the governments of Mexico, Mozambique, Zimbabwe and the Philippines. In 1985 WIIO awarded his Hesperian Foundation (which he founded in 1973) group of workers its first international award for education in primary health care. Despite this renown he and his co-workers:

remain determinedly simple in themselves. Their personal incomes and professional budgets both are nickel-and-dime affairs. To make the books they publish as available to as many people as possible, they keep the prices as low as possible. And they allow others to reprint their work provided the reprints are not sold at a profit.

(Allen op. cit., p.184–5)

In 1987 the Hesperian Foundation had a budget of some US$350,000 from various charitable sources, including Oxfam, UNICEF, SIDA and a number of US foundations. Their main projects are Project PROJIMO (a villager-run rehabilitation programme for disabled children) and Project Piaxtla (a villager-run health-care network), both in the mountains of western Mexico. The Foundation publishes an occasional, fascinating and down-to-earth newsletter.

Despite Werner's advocacy of 'informed self-care', and the need for its institutional support, he is well aware of the limitations of self-care under conditions of oppressive power relations, and of why

169

appropriate primary health care is so often not forthcoming. Even as *Where There Is No Doctor* was being published, he wrote:

> The great variation in range and type of skills performed by village health workers in different programs has less to do with the personal potentials, local conditions or available funding than it has to do with the preconceived attitudes and biases of health program planners, consultants and instructors. . . . Health care will only become equitable when the primary health worker takes the lead, so that the doctor is on tap, and not on *top*.
>
> (Werner 1977b, pp. 4, 12)

However, even if the experts could be brought 'on tap', the achievement of health for all will require other fundamental changes.

> In today's world the biggest obstacles to 'health for all' are not technical, but rather social and political. Widespread hunger and poor health do not result from total scarcity of resources, or from overpopulation, as was once thought. Rather, they result from unfair distribution: of land, resources, knowledge, and power . . . too much in the hands of too few.
>
> (Werner 1985, p. 1)

In today's world there are a number of giant profit-making ventures that are taking an extraordinary toll on the health and lives of billions of people, and that have an enormous negative impact on the well-being and survival of children. These health-destroying multinational industries include:

- alcoholic beverages
- tobacco
- illicit narcotics
- pesticides
- infant formula (powdered milk)
- unnecessary, dangerous, overpriced pharmaceuticals
- arms and military equipment
- international money lending.

Their cost in terms of human life and health is inestimable. The weakened resistance – physical, economic, mental, and social – caused by these unscrupulous businesses adds enormously to the impact of infection and malnutrition. And, as usual, it is the poor who bear the brunt of the damage – especially since the alcohol, tobacco, pesticide, infant formula, pharmaceutical, arms and banking industries have all increasingly targeted the Third World as their new, most vulnerable market.

Attempts have been made by non-governmental organizations, the UN, and the governments of various countries to try to limit the damage caused by these powerful industries. But in the case of each and every one of these killer industries, the US government has defended their interests at the expense of the health, quality of life and often survival of millions.

(Werner 1990, pp. 4, 5)

Later in the same paper Werner cites an international study carried out by the Rockefeller Foundation which concluded that the four key social factors underlying good public health were:

1. political and social commitment to equity;
2. equitable distribution and access to public health care;
3. uniform access to the educational system with a focus on the primary level; and
4. availability of adequate nutrition at all levels of society in a manner that does not inhibit indigenous agricultural activity.

(ibid. p. 7)

Werner has written of his work:

My books and writings on people-centered approaches to health and development are having an increasingly widespread impact around the world. *Where There Is No Doctor* is now in over 40 languages. The most recent translations, completed in May and June, are in two of the indigenous languages of Ecuador. To me one of the most exciting developments with these books has been the collaborative decision by the Ministry of Health and the Ministry of Education in Mozambique to place a copy of the newly revised Portuguese edition of *Where There Is No Doctor* in all of the primary schools of the country, together with guidelines in its use. SIDA (the Swedish International Development Association) is helping to cover the costs of the 5,000 copies of the book. These books and guidelines will allow primary school teachers in the thousands of rural communities that have no health centers or trained health workers to work directly with students and their families in meeting some of the most basic health needs, such as control

171

of diarrhoea through oral rehydration. This plan has the added benefit of making education more relevant to the immediate lives and needs of the school children and their parents, which has been one of Mozambique's goals since liberation.

My own concern has focused increasingly on the socio-political changes that are needed in industrialized nations – especially the United States – if the dispossessed and hungry of the world are ever to be given equal rights and a fair share of the world's resources.

(personal communication 21 May 1989)

Gonoshasthaya Kendra (Bangladesh)

In 1971 a 29-year-old Bangladeshi doctor, Zafrullah Chowdhury, undergoing post graduate training in surgery in London, returned immediately to Bangladesh to play a part in the liberation war against Pakistan, setting up a 480-bed field hospital for the wounded freedom fighters and refugees from Bangladesh near the Indian border. On cessation of hostilities in 1972 he and some medical colleagues established Gonoshasthaya Kendra (GK – The People's Health Centre) 40 km north of Dhaka. GK has from the beginning emphasised independent, self-reliant and people-orientated development. Working originally in the health field it has steadily expanded the scope of its work into additional important areas that affect the lives of the majority of rural-based Bangladeshis: education, nutrition, agriculture, employment-generation, production of basic medicines, women's emancipation. But, it is in the health field that its work has been most innovative. It was the first place outside China to train paramedics seriously and its experience (and Chowdhury's) were important in drawing up the WHO's Alma Ata declaration on 'Health for All' in 1978. In 1989 160 paramedics covered a population of 180,000 from the GK centre and ten sub-centres which have been established round the country. The paramedics are trained in preventive medicine of all kinds, simple curative medicine and operate a health insurance scheme based on ability to pay. They also give advice and practical help on agricultural matters and opportunities for income generation. The great majority of paramedics are women. Their achievement in health service delivery is noteworthy. Infant mortality rate (IMR) in the Savar project area (population 129,000) came down to 60 while national

IMR is 140, per 1,000 live births; similarly maternal mortality rate (MMR) to below 2 while national MMR is between 5.7 to 6.4 per 1,000 births.

In 1981 GK did what it is most famous for: set up the company and factory of Gonoshasthaya Pharmaceuticals Limited (GPL) to make essential drugs of the highest quality at low cost with the additional objective of utilising half the profits for GK's social projects while the other half would be reinvested into GPL itself. It has been an unqualified success and now supplies an average of 15 per cent of all Bangladesh's drugs, but as much as 80 per cent of some sorts. Just as importantly, the fact that in 1981 when GPL was set up its prices were up to 60 per cent below those of the multi-nationals has meant that prices generally have fallen greatly. The factory employs some 400 people, most of whom are rural poor women to whom GK gave functional education and practical training before employment.

The GPL experience meant that Chowdhury was a key adviser to the Bangladesh government in 1982 when it drew up its Rational Drugs Policy and Drugs (Control) Ordnance, proscribing 1,700 dangerous or useless drugs, on the basis of sixteen detailed criteria on use, benefit and harm. Under colossal pressure from foreign governments and multi-nationals, the proposals were watered down somewhat, but were still passed and set a unique example to other countries of how to control their drugs market. It is an example that has been followed by over one hundred countries, including industrial countries like Britain, Germany and France, which have prepared their own essential drugs lists, while over forty countries have prepared national drugs policies based on these lists.

Still in the health field, GK publishes a monthly health magazine with a circulation of 25,000. It has set up a People's Health Herbal Medicinal Plant Research Unit which employs eleven people. It is manufacturing the raw materials for antibiotics and is about to start producing homeopathic medicines. It is negotiating with France for the transfer of vaccine-production technology. And the go-ahead has been given by the government for the establishment of an Institute of Health Science that will train doctors specifically in community health and medicine relevant to Bangladesh's needs, as opposed to those of the West.

In education, GK has an extensive programme encompassing, apart from the paramedics, literacy for all ages but especially children, and training of various sorts. The education is both practical and across many different everyday topics. One innovation is that the children who come to the school are expected to give classes to other children back in the villages. One hundred and sixty three people work in GK's schools.

GK has always placed a special emphasis on women's development and has sought to enable women to break out of the stereotyped female occupations. Different income-earning initiatives for women include metal fabrication, brick-making, block printing, the manufacture of jute and fibreglass, woodworking, shoemaking, baking and weaving. These projects employ 250 women. GK has launched an irrigation project in which the water-pump operators are women. A textile mill is planned. Some agricultural work is compulsory for all GK staff (including Chowdhury) to underline its importance. There is a model farm at the Centre and extensions round the country, backed up by a rural credit scheme which serves over one hundred villages.

GK is controlled by a charitable trust, the GK Trust, of which Chowdhury is one of four trustees. He is also the Chairman of GP. The Trust now employs in total some 1,200 people full time, with 1,000 part-time workers. Its budget is Tk220 million (US$6 million) of which about 50 per cent is self-generated.

An important principle is that, except to the absolutely indigent, GK never gives away its goods and services. They have to be paid for, however cheaply. This explains the relatively high self-sufficiency rate. 'Payment for services helps maintain a feeling of self-respect and self-reliance. . . . The very poor receive a similar level of service to the rich, but it's a service they have paid for' (Chowdhury 1989).

George Vithoulkas (Greece)

The Greek Encyclopedia *Papyros Larousse Britannica* (vol xv, p.396) calls George Vithoulkas 'the leading homeopath of the world and one of the most well-known exponents of homeopathy that has done the most to upgrade homeopathy in the twentieth century' (quoted in ASHM undated, p.1).

Vithoulkas was born in Athens in 1932 and took up homeopathy after taking a degree in civil engineering. He received a degree from the Indian Institute of Homeopathy in 1966.

Returning to Greece in 1967, Vithoulkas started practising and teaching homeopathy to a small group of Greek M.D.'s. The therapeutic success of these first doctors so attracted the attention of others, that the Athenian School of Homeopathic Medicine was established in 1970. The School is devoted to the teaching of M.D.'s exclusively. One year later, in 1971, following an increasing interest on the part of physicians, the first Greek homeopathic society was established, under the name 'Hellenic Homeopathic Medical Society'. In 1972, Vithoulkas started the first Greek homeopathic journal 'Homeopathic Medicine'. In 1976, he organized the first Greek international homeopathic seminar. Since then, seminars, attended by health professionals from at least 20 countries, are held every year.

Vithoulkas' books, *Homeopathy: Medicine of the New Man*, written for laymen, and *The Science of Homeopathy*, directed toward health professionals, have been translated into ten languages and have had a profound influence upon the acceptance and practice of homeopathy worldwide.

Currently, as director of the Athenian School of Homeopathic Medicine, George Vithoulkas heads a team of 30 doctors who practice homeopathy while they study under his supervision. He has established homeopathy in Greece as a science respected by the medical profession, and also made his country one of the leading centers for homeopathy in the western world.

(ibid. p.5)

George Vithoulkas' overall vision is the establishment of homeopathy on a worldwide basis. A major goal in this vision is the establishment of homeopathic medical colleges in the USA and Europe where homeopathy can be taught at the highest level. This will require the development of teaching materials, including books by Vithoulkas and audio-videos for live-case presentation, the training of teachers of homeopathy and the establishment of places for homeopathic education. Organisations associated with Vithoulkas which have the potential to be developed into teaching institutions currently exist in Greece, the US, the UK, Belgium and Norway. Twice a year in Celle, Germany, Vithoulkas also teaches an international group of medical doctors and health practitioners from all over the world.

At present Vithoulkas is spending most of his time writing a new Materia Medica 'including all the contemporary knowledge and my own experience from cases which are mostly taken from the Centre in Athens where we have treated over 150,000 cases' (personal communication, 16 February 1988). The first volume, already completed, will appear in English, German and Dutch. There will be six volumes in all which Vithoulkas plans to complete in six years. Other productions are the Vithoulkas Expert System computer program, which is now on the market and has been 'greeted with enthusiasm by homeopaths' (personal communication, 18 April 1989) and a series of video courses made from his courses. His book *AIDS; The Real Cause and a New Model of Health and Disease* is also soon to be published in English, German, Dutch and Italian.

Geshundheit Institute (USA)

The Gesundheit Institute (GI) was founded in 1971 by Hunter (Patch) Adams, MD, then 26 years old, who described it thus in a 1987 publication of the Institute:

> GI is an experiment in holism with a medical focus based on the belief that one cannot separate the health of the individual from the health of the family, the community and the world. Gesundheit is a socio-political act that grows out of a deep concern for the quality of people's lives in a world dominated by the values inherent in greed and power. This experiment has gone to extremes to provide an active model for what human potential can be if people work and play cooperatively and interdependently. We have taken one of the most expensive things in America, medical care, and given it away for free.
>
> (GI 1987, p.2)

In an interview with the *Washington Post* newspaper (14 February 1988) Adams summed up what was wrong with medicine in the US:

> My wife and I saw four major issues that needed to be addressed: the cost of care, the dehumanization of care, malpractice and third-party reimbursement. It was easy to address them: don't charge money, see the ideal doctor–patient relationship as friendship, don't carry malpractice insurance,

and don't accept third-party insurance. So Gesundheit never accepted money.

(Adams 1988, p.22)

The money came through donations from well-wishers, part-time jobs of the staff (which in 1983 numbered twenty medical workers including three MDs), and Adams' lectures and 'wellness shows'. He is an accomplished theatrical comic as well as an experienced event-organiser, networker and bibliophile (with a 10,000 volume library), movie maker of sixty hours of movies on health, education, children and humour, and a citizen-diplomat, who has been on six peace-making visits to the USSR. As for medicine, by 1983 15,000 patients had passed through the community-style house which was the Institute's HQ and home to key staff.

In 1983 Gesundheit stopped seeing patients to concentrate on raising money for and building its next stage: a forty-bed hospital in a holistic health community on a 310-acre site in West Virginia.

> Much more than simply a medical center, the Gesundheit facility in West Virginia will be a microcosm of life, integrating farming, arts and crafts, performing arts, education, nature, friendship and fun with medical care. We will be a happy, funny, loving, cooperative and creative community – owned by no one, serviced by all, for all people. Family and community structures have broken down to such an extent that we feel the need to provide an opportunity for people to experience them in a healthy context. We want to provide an environment for learning the skills necessary to make community and family live in each person's life. We have seen the impact this can have on individuals both medically and socially in the 20 years we have been operating and deeply involved with at least 15,000 patients using our home as a health facility.

(GI 1987, p.2)

So far about US$900,000 have been raised and the first big building built, mainly with volunteer labour, with estimates ranging from US$4–5 million still needed. Adams thinks it will take four years to complete.

Adams elaborated further the Gesundheit medical philosophy in the *Washington Post* interview already cited:

> At Geshundheit, we see deep, intimate friendship between patient and doctor as having great medicinal power. By

177

being free of charge, we enhance the potential for forming that relationship. When a person comes to me, the first goal is to have a friendship happen out of that relationship. So we spend three to four hours in the first meeting. By the end of that time, I hope we have a trust, a friendship starting to develop and from there we can proceed.

We do not carry malpractice insurance. . . . Malpractice insurance sets up an adversarial relationship with your patients. It eliminates intuition, it forces you to prescribe the cook-book treatment rather than something else, even if the doctor feels the accepted treatment is inadequate or damaging. Where is there room for creativity?

The whole malpractice thing inadvertently reinforces the doctor-as-God concept. If we can't make mistakes, we must be perfect. It also implies that the doctor is responsible for the cure and the patient the passive recipient of it. The huge majority of illnesses have a lifestyle component – ultimately, the health of each of us is our responsibility. We'll have a sign in our hospital that says 'Please live healthy – medicine is an imperfect science'.

Early on, we realized how ill-equipped allopathic medicine, which is what an MD practices, is to deal with all health needs. So within the first year, we started letting an acupuncturist come and work with us. Since then, we have been very open to naturopathy, homeopathy, acupuncture, chiropractic, lots of things. We avoid the arrogance of thinking that any particular system will help more than a fraction of the people coming for help.

From the start, it was obvious to me that we had to have fun in what we were doing or the staff would have left in a week. Forget the patient, it had to be fun for us. Life has to be fun! Not only is fun a glue for our community, but it had overwhelming medicinal effects on the patients. So many fewer pain medications! With psychiatric patients, there was overwhelming progress in intimacy and in relieving symptoms. Health is typically defined as the absence of disease. To me, health is a happy, vibrant exuberant life every single day of your life. Anything less is a certain amount of disease.

When we open in West Virginia, we will have some doctors who don't do four-hour interviews. I am trying to hire a staff with broad points of view, so that whoever walks in there will be able to go to somebody that they can be touched by.

178

As for doctors, I just put together a booklet of 100 letters we have gotten, which is just 1 percent of the letters we have received, from doctors, nurses, medical students, in essence describing how much medical practice is painful to them, and how much they are behind us, how much they would want to work with us, how much we represent the little hope that they have in medicine and why they are not quitting medicine. All of them, doctors, nurses, just crying out for a meaningful context in which to practice medicine.

(op. cit. pp.22, 23)

Dr Aklilu Lemma and Dr Legesse Wolde-Yohannes (Ethiopia)

For twenty-five years two Ethiopian scientists have been struggling to win international recognition and support for their discovery of a cheap, community-based preventive of a chronic and destructive Third World disease. The disease is bilharzia, or schistosomiasis, a debilitating and eventually fatal illness caused by flatworm infestation of the liver and other body organs, which afflicts more than 200 million people in seventy-four countries of Africa, Asia and Latin America. Present molluscicides, to kill the snail-carriers of the disease, and therapies for bilharzia, are far too expensive for the communities that need them.

Lemma's discovery, in 1964, was that the fruit of a common African plant, the endod or soapberry, acts as such a molluscicide. The endod berry has been used for centuries by African women as a soap to wash their clothes. Lemma made his discovery by observing a pattern of dead snails in a river downstream from where some women were doing their washing.

To explore the implications of his discovery, Lemma established in 1966 the Institute of Pathobiology in Addis Ababa University and for the next ten years directed a team to undertake systematic research on endod as a molluscicide. He was joined in this work in 1974 by Wolde-Yohannes, who continues with endod research there to this day.

Given the immense promise of the discovery, confirmed by the early research, progress in making this molluscicide available to the people who so desperately need it has been tragically slow. The reasons for this expose some of the least attractive biases and institutional failings of the international medical community:

179

- Third World research is still not taken seriously in the West and does not qualify products for international acceptance. Lemma's early work at the Institute of Pathobiology therefore failed to win the recognition it deserved. Dr R.M. Parkhurst, Senior Organic Chemist at Stanford Research Institute, USA, who has also worked on endod, has said: 'I think the toxicology has been done already. It just hasn't been done by the "recognised source". People just don't want to believe anything that comes out of a Third World university and that's just too bad' (quoted in Steele 1987, p.72).
- Western research is very expensive but without it international (UN) organisations will not promote a product. Even most Third World governments can or will not proceed with large-scale dissemination of a new remedy without the support of such organisations.
- Endod offers the prospect of being a poor country's solution to a poor person's disease. But endod's very cheapness and simplicity mean that its commercial possibilities are not attractive enough for private enterprises to invest the necessary resources to win toxicological clearance.
- Mobilising public money on the necessary scale has been problematic and time-consuming, with a curious reluctance from key institutions, especially the World Health Organization, to commit themselves to the development of endod's beneficial properties.

In the last few years, Lemma's and Wolde-Yohannes' persistence seems to have made progress towards overcoming these problems. The support of key scientists and donors in the West has opened the doors to the necessary laboratory and field trials. Much of the work will simply repeat the early Ethiopian research, but this time under so-called internationally-recognised 'good laboratory practice'. This means that the large sums of money raised will do little to build the local institutional and research capabilities of the Third World country which actually did the pioneering work, but will simply bolster established centres in the West. If the Third World ever reaps the health benefits of endod, it will be due not just to Lemma's and Wolde-Yohannes' scientific skills, but also to their stubborn refusal to submit to an international medical establishment that has systematically sought to ignore or devalue their work.

180

EDUCATION

As in health, so in education Western models have now come to dominate the world and are now applied in very inappropriate situations. The two cases to be explored here deal with new approaches to schooling in Third World countries.

Foundation for Education with Production (Botswana)

The Foundation for Education with Production was established in 1980 by Patrick van Rensburg, a liberal exile from South Africa, who had settled in Botswana in 1962. There he founded the Swaneng Hill School and, following its success, two other schools in association with the Botswana government; the Swaneng Consumers' Cooperative; and the Brigades' Movement. It was these experiences through the 1970s that led to his creation of the Foundation.

Van Rensburg's educational approach was radically different from usual practice. The curricula had a strongly practical orientation, including agriculture, building, typing, for example, as well as giving more traditional school instruction. While his teaching standards were as high as anywhere in Botswana, the schools were low cost because of the traditional, frugal living standards they embraced. And the cost was further lowered, in an effort to bring the schools within reach of ordinary Botswanans, by the Brigades, which were self-help organisations of the students which produced goods and services both for the schools and for sale to help finance their education.

In building on these early experiences, the Foundation for Education with Production sought to create a new blend of theory and practice in education, to be spread internationally with and through autonomous national committees. It set up education with production projects in Zambia and Zimbabwe. It devised and held workshops on a wide range of subjects including development studies, environmental and social studies, co-operatives, fundamentals of production and many others. It organised conferences and seminars in Grenada, Botswana, Zimbabwe, with Southern African liberation movements and in the UK. It trained people to run courses on education with production. It published a directory of projects, newsletter and a journal. And it engaged in research. All these activities are rooted in the Foundation's core perception of education-with-production as a cornerstone of community development.

An example of education-with-production in practice comes from Zimbabwe, where education receives 22 per cent of the government budget, the fifth highest in the world, and education-with-production is an officially sponsored approach. 'At Tafara in Mabvuku district schools pupils dug a fishpond in the shape of a world map' thus learning:

> not only about fish farming, but also about geography, building techniques, food and nutrition, the environment, market economics, book-keeping and the control of water-related diseases such as malaria. . . . All schools now have fruit and vegetable gardens. Sahumani School in Manicaland, for example, earns $500 a year from the sale of its produce. Plots of eucalyptus trees planted by the school also provide the community with firewood and help to prevent erosion.
>
> (UNICEF 1990, p.58)

Van Rensburg has described the concept of education with production thus:

> Education with Production is a key component of a progressive and non-elitist education, one which links theory and practice, which links school and community and is essentially a lifelong and whole social process.
>
> (Van Rensburg 1984, p.6)

In this new conception of education, people should develop themselves mentally as fully as they can, have all modern knowledge available to them, but also have technical skills. They should understand the processes of production and how society has developed. Linking education with production is part of the whole process of social, economic and political development itself. The more we can involve the ordinary people, who themselves then understand the processes and are able to continue with them, the greater chance we have of making sure that we create alternatives which will feed people, clothe people, house people and ensure their good health. All these strands of development in which they are involved are essentially the educational process.

(Van Rensburg 1981, p.8)

New School Programme (Colombia)

The New School Programme (NSP) was designed in 1975 by a team of rural educators under the leadership of Vicky Colbert, a Colombian educationalist who has been involved with it ever since. From 1983–5 she was Deputy Minister of Education; now she is a Senior Education Adviser to UNICEF for Latin America and the Caribbean.

> The New School is defined as a system of primary education that integrates curricular, community, administrative-financial and training strategies; and allows for the provision of complete primary education as well as a qualitative improvement in the nation's rural schools.
>
> Essentially, this system provides for active instruction, a stronger relationship between the school and the community, and a flexible promotion system adapted to the lifestyle of the rural child.
>
> (Colbert 1987, p.14)

The principal characteristics of the NSP approach are summarised as follows:

- *Active learning* – students have their own materials; a small library and self-instruction workbooks are furnished free by the government. They study in small groups and learn to solve their problems by doing and playing. Children learn to be active, responsible and to participate.
- It offers a *multigrade approach* that permits complete provision of primary values where incomplete schooling exists.
- It includes its *own school government* through which children are initiated in civic and democratic attitudes. Habits of co-operation, comradeship, solidarity and participation are encouraged. Children learn to act responsibly in the organisation and administration of school.
- The design of the *study guide* takes into consideration the learning pace of each child. Promotion to the next objective or grade is flexible. There is no repetition.
- Materials are *affordable*, because one set is for two or three children and they last for several years. They have core and local components, are for individual and group work combined.

183

They link school with community and integrate cognitive with social development.

- The classroom is a *dynamic work area* with a library and interest areas or learning centres to support the activities developed by children and community.
- The school operates as the *centre of information* and as an integrating force for the community. Parents participate in school activities and the school promotes actions that benefit the community.
- Teachers are *facilitators*. They guide, orient and evaluate learning. They are trained in workshops that follow methodologies similar to those they will later apply to their pupils.

The World Bank selected the New School as one of the three best experiences in the world for successfully applying broad innovative education techniques in primary schools. The Bank states:

We believe that the experience with regard to implementation of this programme is highly instructive since it contains a number of important lessons that merit broad diffusion among educational planners and policy makers in the developing world as well as among the staff of international organisations that support educational development.

(personal communication, 19 November 1990)

The New School Programme was also presented in the world conference Education for All as one of the two roundtables from Latin America. NSP's worldwide relevance has already been recognised by UNESCO and UNICEF, as well as by the World Bank, and the governments of over fifteen countries have sent representatives to Colombia to learn from it.

The impact of the New School Programme has been widespread: '(It) is being implemented by 35,000 coordinators and 26,000 teachers in 15,000 schools to reach 750,000 children a year. . . . A major political, technical and financial decision has been made in order to reach the entire 28,000 rural schools by 1992' (Colbert 1987, p.24).

In addition, Vicky Colbert has organised an education foundation with support from the Inter-American Foundation to continue innovating and introducing qualitative improvements in the New School methodology. An adaptation to urban marginal areas is one of the projects that has been initiated with promising prospects.

HOUSING

Whenever the word is used, 'housing' tends to be interpreted as a noun – meaning shelters that are built. Who does the building, for whom, and how, i.e. the subject, object and process of the word 'housing' used as a verb, tend to be secondary considerations, the usual assumption concerning which being that the subject will be either the market or the state, the object will be largely unidentified 'people' and the process will be one of minimal interaction between the two through the impersonal mechanisms of bureaucracy or price.

The man who has, perhaps, done most to challenge that assumption, first by pointing out that most houses in the world are, in fact, built by people for themselves; and second by stressing the efficiency of the community-based organisation through which people can build for themselves on a large scale, is John F. Charlewood Turner. Of Turner's work, William A. Doebele, Frank Backus Williams Professor of Urban Planning and Design at Harvard University, has written:

> John (FC) Turner's approach, based on self-reliance and the traditional wisdom of mankind, has touched the policies of virtually every national and international organization in the world. It is literally true that millions of families and tens of millions of individuals have gained both better shelter and, more importantly, self-respect as a result of his tireless advocacy on their behalf.
>
> (personal communication, 31 May 1988)

John F. Charlewood Turner (UK)

John F. Charlewood Turner was born in England in 1927. He graduated in architecture from the Architectural Association in London in 1954 and worked in Peru from 1957 to 1965, mainly on the advocacy and design of community action and self-help programmes in villages and urban squatter settlements. From 1965 to 1967, he was a Research Associate in Cambridge, USA, at the Joint Center for Urban Studies of the Massachusetts Institute of

Technology (MIT) and Harvard University and then lectured at MIT until 1973. Returning to London, he was a lecturer at the Architectural Association and the Development Planning Unit, University College London, until 1983, when he resigned to devote himself full time to his non-profit consultancy AHAS, which he established in 1978 with his wife and co-director, Bertha, and two other colleagues.

During these years, Turner's many publications have had a great influence on housing policies worldwide. They include: *Freedom to Build: Dweller Control of the Housing Process*, (with Robert Fichter; Macmillan, 1972), translated into Italian and Spanish; and *Housing By People: Towards Autonomy in Building Environments*, (Marion Boyars, 1976), translated into German, Spanish, French, Dutch and Italian.

From 1983 to 1986, Turner was co-ordinator of Habitat International Coalition's (HIC) NGO Habitat Project for the United Nations International Year of Shelter for the Homeless (1987). The HIC Project carried out a global survey of recent and current local initiatives for home and neighbourhood improvement. From over 200 Third World cases identified (from a total of 340 cases worldwide) twenty were selected for in-depth documentation by local researchers in collaboration with the communities concerned. Summaries of these cases provide the core of *Building Community: A Third World Case Book*, edited by Bertha Turner (Turner, B. 1988) and with an Introduction and Conclusions by John F. C. Turner.

Turner's frequently quoted conclusion from his own field experience was first published in *Freedom to Build* (Turner and Fichter 1972):

> When dwellers control the major decisions and are free to make their own contributions in the design, construction, or management of their housing, both this process and the environment produced stimulate individual and social well-being. When people have no control over nor responsibility for key decisions in the housing process, on the other hand, dwelling environments may instead become a barrier to personal fulfilment and a burden on the economy.

(p.241)

In his latest work Turner has tended to stress the concept of *enablement* as much as freedom:

186

Enablement is the key. Neither bureaucratic mass housing nor the uncontrolled market can build communities and eliminate homelessness. But *people* can, when they have access to essential resources and when they are free to use their own capabilities in locally-appropriate ways . . .

The fact that so many people have done so much with so little in low-income countries, while so little is done for low-income people with so much by their governments, demonstrates the necessity of the radical policy changes which are already taking place. Increasingly, with some and perhaps vital assistance and encouragement from international agencies and NGO's, Third World government policies are changing over from vain attempts to supply public housing to the support of locally self-managed initiatives. The necessity of enabling policies is not so obvious in countries whose governments can afford to subsidise all who cannot pay current market prices. But as the longer-term social and economic costs of depriving people of their freedom of choice and responsibilities turn people's demands to be housed into demands to house themselves, we become increasingly interested in Third World experience and what it can teach.

(Turner, B. 1988, pp.14,16)

The Turners are now organising and making more accessible to others the mass of documentary material relevant to enablement of local communities. This process will prepare the ground for the sharing of information for promoting enabling policies and the development of better tools for understanding and practising community-building.

Turner's emphasis on the ability of people to build their own communities with their own organisations, and the efficiency with which they can do this, has now become conventional wisdom. However, far less often recognised is the essential role of both the market and the state if community organisations are to realise their potential:

Sustainable development is possible only when local, com-munity-based initiative is supported and enabled by the state and the market.

Responsible and creative power over the use of resources depends on knowledge and facts the visibility of which

187

depends on the viewpoint. From their own personal and local viewpoints, people have unique access to knowledge of their own situations and priorities. Looking outwards and upwards from local situations, however, makes it difficult to see connections with other situations and localities. On the other hand, policy-makers and administrators, specialists and managers looking down at their constituencies or markets can see the connections but they cannot cope with the infinite variety of personal and local realities.

In other words, the three powers have different and complementary potentials and limitations. State powers are the only ones that can maintain authority over the basic institutions: the law, the exchange system and the structure of government. An equilibrium of supply and demand cannot be achieved without market forces. Neither central authority nor corporate commerce can be economically sustained without community-based initiative.

Worldwide experience of home and neighbourhood building and management reflects the overall pattern. The activity of housing, housing understood as a verb, is a paradigm for human life in the world. This vital activity demands more living space, lifetime and energy than all other activities together. The facts show that when housing and local development are community-based, supported by state powers and served by market forces, far more is achieved with far less. Development then means human fulfilment.

(Turner, J.F.C. 1988, p.1)

Orangi Pilot Projects (Pakistan)

Orangi, situated on the periphery of Karachi, is one of the largest squatter settlements in Asia, with a population close to one million and covering 8,000 acres. It is still growing. Orangi residents:

depend mainly on 'informal' (often underground) sources. Land is obtained through *dalals* (touts); credit, materials and advice from *thallawalas* (block manufacturers). Self-supporting private schools and coaching centres teach their children; private doctors and quacks (physical and spiritual) treat their ailments. They continually resort to the black market and the bribe market for business facilities or welfare amenities or peace from harrassment.

(OPP 1988, p.3)

Through these 'informal' sources Orangi residents have been intensely productive. They have laid 6,350 lanes and built 95,000 houses without any help from the Karachi development authorities. They have established over 500 schools with over 50,000 pupils without any assistance from the Directorate of Education. Most of them earn their living through private family businesses set up with their own savings (Khan 1990a, pp.1, 2).

However, the houses in particular suffered from severe problems due to poor design, defective building materials and lack of sanitation. Into this environment in 1980 came Dr Akhtar Hameed Khan, a well-known social scientist from Bangladesh, at the request of one of Pakistan's leading bankers, Aga Hasan Abidi. Khan decided to develop a research project aimed at development through community organisation.

> We are all living through a period of social dislocation. There is a need to re-establish a sense of belonging, the community feeling, the conventions of mutual help and cooperative action, that can be done chiefly through the creation of many kinds of organisations, social and economic. Without such organisations, chaos and confusion will prevail. On the other hand, if social and economic organisations grow and become strong, services and material conditions, like sanitation, schools, clinics, training and employment will also begin to improve.
>
> (ibid. p.2)

From the start OPP functioned as a research institution, analysing problems in Orangi and then, with the residents, identifying solutions. The actual implementation of the solutions always lay with the residents themselves. OPP's role was limited to bringing them together, motivating them to organise, and giving technical and other professional advice.

OPP's first, and most spectacularly successful, project was the provision of low cost sanitation, begun in 1981. This was intended to address the appalling drainage and sewerage problems in Orangi which resulted in frequent severe flooding and innumerable health problems. OPP's approach was to encourage the lanes in Orangi, comprising twenty to forty households, to form their own organisation, whereupon OPP would advise them how they could best construct underground sewerage for the whole lane. The works would be paid for and managed by the lane organisation, i.e. the residents, itself.

Needless to say, such organisations cut across many local political and other vested interests in Orangi but, once a few lanes had overcome these, the benefits were too obvious for the vested interests to be able to resist them, and the project spread very rapidly. By September 1990 4,396 lanes had laid self-financed and self-managed underground sewerage lines, involving 67,670 houses and investment by the people of some US$2.27 million. This compares with another project elsewhere in Orangi, which was organised and administered by the UN Committee on Human Settlements without popular participation, on which US$650,000 of external aid was spent on developing thirty-five lanes and 552 homes over three years (Hasan 1986). These figures make OPP twenty-six times as cost-effective as the conventional UN programme.

On the basis of this early success, OPP has established several other programmes:

- A low cost housing programme,
- A basic health & family planning programme,
- A women's work centres programme,
- A programme of supervised credit,
- A schools programme.

(Khan 1990b, pp.1, 2)

OPP's housing programme has so far concentrated on research and in four years has managed to develop a design that can be built by local masons with traditional materials; that has walls eight times as robust; that has much improved roofing and proper ventilation and sanitation; and that costs one third of the usual construction. OPP's forty demonstration units are now the subject of intense interest locally from low-income home owners.

The health and family planning programme began with a pilot project involving 3,000 families and operating through seventy-five home activists each of whom had educational meetings with 25–30 housewives. The package of advice and services covered training about causes and prevention of common Orangi diseases; immunisation services; family planning training; delivery of family planning supplies and services; kitchen garden advice and service; advice on nutrition and child care. Preliminary results of a survey in the first three months of 1989 indicated that 44 per cent of those in the programme practised birth control (compared to a 9 per cent national average) and 90 per cent had immunised their children (OPP 1989).

Following this success the Programme has been formally registered with the authorities as a social welfare association, seeking to train in these subjects eighty home activists a year who would in turn pass on their knowledge to some two to three thousand women annually (OPP 1990, p.12).

The Women's Work Centres (WWCs) Programme was begun in 1984 and after many difficulties became self-managing and self-financing by 1989. OPP focused on the large numbers of stitchers in Orangi sewing simple textiles for export, often under very exploitative conditions. It set up a trust which had two roles: to procure export orders and set up and train work centres in Orangi homes to fulfil them punctually and to the requisite quality. OPP stressed from the outset the need for the centres to become commercially viable. In 1989 the WWCs earned their stitchers nearly Rs2.5 million at a rate of 30 per cent above their old contractors' wages. The OPP Trust had withdrawn entirely from support and advice and only gave loans for the purchase of new equipment. New family work centres were being set up spontaneously. A self-sustaining process was in train.

In 1987 OPP sought to extend its work centres with stitchers to the support of other enterprises through a programme of credit for family enterprises. Some 400 units had been financed to 1990, with mixed but generally encouraging results.

OPP's research and extension approach is conceptually simple and wide-ranging in application. Participatory research analyses the problem and finds an appropriate solution; extension then informs, motivates and enables people to implement the solution themselves. Hasan's conclusions are telling:

> The Low Cost Sanitation Programme of the OPP has shown that the squatter settlements of Karachi can build their sewerage system without international aid, without waiting for the local bodies to take decisions (usually the wrong ones) and without catering to a system of development of which consumption is an integral part. Furthermore, in the process of acquiring these services they can bring about a change in the unequal political relation that they have with those who rule them.
>
> (Hasan 1986, p.18)

This last point is of crucial importance. The OPP approach not only delivers services, it also begins the process of reversing the community's relation to the market and the state so that these institutions begin to serve it rather than the reverse. As this reversal becomes more pronounced, and the market and state become more supportive of the community endeavour, there is every prospect that this will become even more productive, to the incalculable benefit of the people concerned.

The Lightmoor Project (UK)

Exactly the same impulses towards self-help and community as operate in Orangi township can be found on a far smaller scale in the British Midlands: the Lightmoor Project on the outskirts of Telford New Town. Here fourteen houses are being painstakingly built by the families which plan to live in them, by a process that goes far beyond erecting structures and is consciously creating community.

The Lightmoor Project was the brainchild of the Town and Country Planning Association and, especially, of its Development Officer Dr Tony Gibson, described by a local newspaper as 'a planner, visionary and ideological descendant of Ebenezer Howard, founder of the garden cities' (Manning 1987). Certainly Lightmoor is a conscious attempt to create the garden city's present-day equivalent, with a contemporary emphasis on community, democracy and participation.

The project began in 1984. Gibson has described its recipe thus:

> The ingredients are land, people and money – marginal land which the commercial developers pass by, a mixture of people, different skills, temperament and backgrounds. But they all have a common interest in allowing each other independence and privacy and the will to work together to decide on things that concern everyone; and the muscle power to turn the talk into action. These two, land and people, attract the money from outside investors, from the building societies who see the point of investing in a neighbourhood that's self-built to last.
>
> (quoted in Knevitt 1987, p.12)

This simple analysis gives little idea of the practical difficulties involved in coaxing such land, people and money together: the twenty-two acres which had to be prised out of the Telford Development Corporation at agricultural rather than housing land prices; the search for the families who wanted to camp in caravans while they laid bricks and mortar, sometimes keeping up an outside job as well; the persuasion necessary to convince a building society to put some £250,000 of its depositors' money into such an unorthodox situation.

Lightmoor's initiative received recognition in 1987 when it was judged 'most outstanding entry' for that year's Community Enterprise Awards, organised by *The Times* and the Royal Institute of British Architects. It is planning a modest expansion in the near future, but it will only be judged a real success by its founders when the model it has pioneered is appropriately replicated in Britain's decaying inner cities whose residents are in direst need of its elixir of hope, community and regeneration. Despite strong verbal backing from the Prince of Wales and rhetorical support for such an approach from politicians of all colours, there is still little sign of this widespread practical extension of the Lightmoor recipe to those who need it most.

The Lightmoor and Orangi Pilot Projects have many common elements: a community-based philosophy; a participatory methodology; a commitment to self-help and self-reliance. Where they differ most, of course, is in their scale. The Orangi Project is being seriously studied as a solution to mass housing problems in the Third World. Lightmoor is still very much the small-scale prototype. Whether it grows larger will depend both on the extent of commitment of public policy and the general level of prosperity. The final irony for Lightmoor is that, despite a waiting list of would-be self-builders in the wings, their land supply is under threat. Telford is now something of a boom town and the land earlier earmarked for Lightmoor is now far from 'marginal'. It is possible that the Development Corporation will not countenance the financial loss incurred in making this land available to Lightmoor for building at agricultural prices. So would fall the first of Tony Gibson's magic ingredients, evaporated by prosperity. Community access to resources is, it seems, conditional on adversity.

SPIRITUAL AND CULTURAL RENEWAL

It will have become apparent during this book that economic development is here perceived to be primarily a cultural rather than a technical process. The last two practical illustrations to be presented here give further strength to this perception. Both involve significant improvements to the local standard of living, but in both cases these improvements come about through commitments to other objectives: to village life and community sustainability in the case of the Finnish Village Action Movement, and to divine service in the case of India's Swadhyaya.

Village Action Movement (Finland)

During the 1960s and early 1970s Finland experienced rural depopulation. By 1974, however, there were the first signs of a modest rural revival based on village committees, which, in 1976, came to the attention of then 39-year old Lauri Hautamaki, an Assistant Professor at the University of Helsinki. Hautamaki was familiar with techniques of action research developed elsewhere and started a project of action research into this embryo movement, working with the municipal authorities and researchers from five other universities to evaluate the potential for revitalisation of rural communities and stimulating the formation of more village committees. In 1978 Hautamaki moved the project with him to the University of Tampere, where he had been appointed Professor of Regional Studies.

The researchers' identification of 'concrete utopias' caught the public imagination and it was the blaze of publicity that followed that caused the Village Action Movement to mushroom.

A book, *Living Village* (Hautamaki 1979) sold 3,000 copies. A series of training broadcasts transmitted by Finnish Radio had 200,000 listeners. Village committees were established. By 1988 some 2,300 such committees had been formed, covering two-thirds of Finland's villagers and initiating projects mainly in the areas of culture, leisure, communications, services, housing and economic development. Because of their knowledge of local resources and potential, the committees are also becoming important new channels for public investment. They involve about 25,000 people directly and positively affect the lives of some 500,000 people (out of a Finnish population of 5 million).

Examining in detail the activities of the committees, it is notable how they amount to an expression of small-scale *collective* action over individualism. Communal facilities are emphasised or restored (e.g. the widespread KOKKE project – 'The School as the Centre of Village Life'), as are public and social services, such as health, postal or transport services.

> Although village activity has achieved much that is of positive value, its importance lies much deeper. In the long run its greatest importance is in the change in people's attitudes. The former passivity and submissivity have given way to a new vigour, self-reliance and community spirit and to better awareness of opportunities for activities and assistance. Such a change in attitude would never have come about simply by providing information and training or by increasing government aid.
>
> Village activity has also done much to influence the change in public opinion in administration and planning. There is pressure for change because of the increasing insecurity in all walks of life and because of the diminishing faith in continuing economic growth. New approaches are being sought in all fields. Planning is changing from being dominated by experts, with the emphasis on big units and directed from the top downwards to a new humanism, which stresses the value of the individual and small, loosely organized units in which confidence is placed in the competence and resources of local people. Self-reliance, individuality, self-sufficiency and the ability of the individual to influence decisions affecting him personally are concepts that continue to feature prominently in discussions being held on the latest (social) trends.
>
> (Siiskonen 1986, p.30)

The village action committees have several striking aspects which have contributed to their success:

- Their vitality and creativity, and their very broad range of activities, together with the flexibility and efficiency of their organisation – the activities encompass the arts (music, drama, painting), crafts (traditional, like weaving, furniture-making, herb-production, printing; and modern, like photography and video), economic development and the encouragement of entre-preneurship, sports, especially winter sports, and social events of all kinds, involving the whole village and generating a palpable enthusiasm and liveliness.

- Their evident knowledge of and love for their village, its culture and traditions and their natural environment.
- Their determined self-reliance in their activities – while many of them have their local government as a partner, their approach seems always to be 'We are going to do this; do you want to help?', which both makes dependency less likely and (ironically) perhaps gets better results, as the municipality is put into a position where it cannot afford *not* to be involved in promising initiatives, rather than the reverse.

Whatever the reasons, the local governments are giving the village committees increasing support. There is also evidence that the committees are reviving and creating a new co-operation between the villages' more traditional organisations (farmers' organisations, trade union branches, youth societies, country womens' clubs, etc.). Household and farming extension services are also reviving, especially, in the latter case, in the field of organic agriculture.

Another significant new idea is the village development corporation, a locally-owned and run, formally-registered body (the village committees are informal) being pioneered by several villages to run public services, identify business opportunities, attract new residents, etc. Several villages are also introducing tele-cottages, making use of the decentralised potential of information technology to enhance village communications and strengthen the local economic base.

Each year a National Village Action Festival brings about 1,000 activists together around a chosen theme and chooses a Village of the Year. In 1988 the village Kaivola was chosen. Its story sheds specific light on the workings of the village movement.

In the 1950s there lived a lot of people in Kaivola village, they had a school with two teachers and 70 pupils, two village shops, a post office and a lot of life. In the beginning of 1970's the signs of death were seen: one of the shops was already closed, people, especially the young ones, were moving away, the number of pupils in the school declined dramatically and the prospects were that the school would be closed in 1982 at the latest.

The initiator of the mobilization in the village was the Association *Suhina* (for the preservation of local culture and nature), which was established in 1959 to restore the old windmill on the highest hill of the village (Suhina means sighing, like the wind in the wheels of the mill). Since 1961 Suhina has been running the summer theater every summer, which draws practically all the people in the village to work voluntarily in different ways to keep the performances going. The income from the theater has made it possible for the village to buy land for the village house and new housing projects to tempt new people to village. Altogether they have so far bought 20 hectares of land.

Kaivola was the first village in Finland which in 1977 completed the village master plan. Three years ago the village road got lights, which was financed by the villagers, Suhina and the township equally. Today there are new houses on 16 of the 23 free plots. The number of inhabitants has increased from 95 in 1976 to 146. There are 39 pupils in the school and next year the number will be over 40, which implies that they will have a third teacher.

One important sign of success is that the retired people don't move away any more from Kaivola, but prefer to build there their new house for retirement. There are a lot of plans for improvement of life in Kaivola in forthcoming years. So far the plans have the time span until year 2050.

(Pietilä 1989, pp.4–5)

Swadhyaya (India)

'Swadhyaya' comes from Sanskrit and can be roughly translated as 'study of the self'. It is the name of a philosophy currently said to motivate into action three million people in up to 100,000 villages in India. Swadhyaya was founded and is led by a 'Maharashtrian saint' of Brahmin upbringing, Pandurang Shastri Athavale, known to his followers as Dada (elder brother).

Born in 1922 into an erudite family, Dada was early immersed in philosophical studies for which he had great aptitude. In 1954 he contributed to the Second World Religion Conference in Japan and returned to India to become a preacher and teacher at the Bhagwad Gita Pathsala in Bombay. It was there that he developed his Swadhyaya approach and encouraged his students to go with him or by themselves into villages to talk with the villagers and actions to spread the philosophy were started.

Transcending the phenomenal self, the 'self' of Swadhyaya
is taken as the manifestation of Divine Being. With this
awareness, individual's sense of isolation and vulnerability is
replaced by self-reverence and identity with others, including
nature, laying the basis for a different order of social creation,
mediated by the love of Divine Being.

(Srivastava 1989, p.1a)

The Swadhyaya philosophy expresses itself in several different ways.
There are the huge meetings addressed by Dada which can number
up to half a million people. His reputation as a preacher seems,
however, to be matched by that for humility and self-effacement
and a very simple personal lifestyle – no tele-evangelist he.

Then there are several socio-economic initiatives based on
Swadhyaya philosophy which would seem to have done much
to improve the condition of the village people, but as byproducts
of their main purpose, which is the service of God and love for
their fellow human beings. For example, there are over 3,500 'farms
for God' averaging five acres, worked by 'Swadhyayees', with any
individual, however, only working a few days a year on any one plot.
Thus they have no personal claim to the wealth produced, which
belongs solely to God and which is either distributed to the needy
or reinvested in more land or farming inputs. The creation of this
'impersonal wealth' is also at the heart of the tree-planting enterprises
of the Swadhyaya 'family'. Thirteen large orchards of fruit trees have
so far been created, each with the involvement of at least ten villages.
Their tending is regarded as an act of worship and the villagers who
do so by rotation are seen as priests in the execution of sacred ritual.

As with farms and orchards, so with fishing communities on the
Maharashtra and Gujarat coast. Eighteen boats have so far been built
and crewed in rotation by local fishers, with the catch again being
regarded as 'impersonal wealth' belonging to God and distributed
to the needy.

Those villages that have become most solidly involved with
Swadhyaya work are encouraged to build temples for worship,
simple structures of local materials and without walls where the
whole village gathers to worship and offer a portion of their earnings
as God's share, in a spirit of gratefulness to Him, for the use of the
needy. Also, social grievances and disputes are raised and resolved
there. These temples acknowledge no distinction or discrimination
between caste, class or gender. Seventy-four of these temples are now

in existence in as many villages. The scope and spread of Swadhyaya make it of immense potential importance and it would seem to have a real potential for addressing the causes of the global crisis at their roots.

For example, it tackles directly the materialism of the Western worldview by reasserting the essential spiritual quality of human nature; it tackles poverty by bringing about increased production but *without* enlisting the greedy, self-serving incentives of the Western economic system, which have so many problems of their own; it establishes a new, harmonious and reverential relationship between people and nature; it involves a fraternity/sorority between all people that transcends distinctions of class, creed, caste, etc. which has, it seems, already been efficacious in the resolution of communal violence in Swadhyayee villages. Swadhyaya appeals to and works directly on the individual self; seeks personal transformation as the basis of communal solidarity and ecological awareness; and roots this transformation every bit as deeply as the subconscious appeal of the Western worldview. This approach would seem to have a real chance of long-term success in establishing a new impulse and motivation for activity and development that foster mutuality, co-operation and care for the unfortunate rather than the jungle ethics of Western competition.

9

CONCLUSION

It should be obvious that the people and projects in this book bear witness to solid, important achievements with immense potential for the future. Bertrand Schneider (1988) calculates that development NGOs currently benefit some 100 million people out of the 2 billion rural poor with whom he is principally concerned, and have brought them great benefits at a cost of only some US$6.50 per head per year (p.213). He then calls for an annual investment of US$13 billion per year to reach the other 1.9 billion people in his target group (p.236). Schneider does not say whether he considers this should be 'new money' or resources transferred from existing allocations. He does make clear, however, that it should be an investment in rural areas in small-scale activities in which the beneficiaries 'retain the initiative, choice and responsibility for development decisions which are aimed at answering their real needs'. Such large-scale projects as may still necessary will then proceed from a quite different motivational, practical and political basis, in which 'the starting point of development efforts is the village or community' (p.226).

A group of South Asian scholars has studied in detail the various processes and conditions by which village-based development can be achieved:

A social transformation of enormous magnitude has to be envisaged. In Third World countries the critical structural changes relate to a shift of decision-making power towards the poor by initiating a 'bottom-up' process, the village becoming the focal point of development, and a change in the education system redirecting it towards raising mass consciousness and remoulding elites. There is in the light of this no easy way to bring about the structural changes required, which themselves have to be supported by an integrated process of total mobilisation, involving raising peoples' consciousness and the inculcation of democratic values, the transformation of labour-power into the means of production, the fullest utilisation of local natural resources and the systematic development of appropriate technology.

(De-Silva *et al.* 1988, p.22)

Furthermore, as one of the scholars quoted above has noted elsewhere:

A truly participatory development process cannot be always generated spontaneously, given the existing power relations at all levels, apathy, and the deep-rooted dependency relationship between rich and poor, common in most countries of the South. It often requires a catalyst or initiator who can break this vicious circle, a new type of activist who will work with the poor, who identifies with the interests of the poor and who has faith in the people.

(Wignaraja 1988)

This 'new type of activist' is a fairly unusual human being: someone with a clear intellectual grasp of social trends and forces, an understanding of commercial and local and national bureaucratic processes, an empathy with and sensitivity to the poor and, usually, a willingness to live on a low income. It is the involvement of such people in NGOs which has propelled those organisations to a position of perceived importance in development work. It is to train, equip and support such people, most of whom will be drawn from the ranks of the poor themselves, that the bulk of Schneider's US$13 billion is required.

The good news is that the sum of money itself is relatively trifling and could certainly be found within existing aid and development budgets if the political will were there, simply by cutting out some of the waste and absurdities identified by Hancock (1989) in current development practice. The bad news is that these absurdities are the predictable consequence of the priorities of highly entrenched vested interests in both North and South which have much to lose from the kind of development advocated by Schneider, Wignaraja and Another Development. The likely result is that people's organisations will continue to turn in highly cost-effective results which are well below both their potential and the need; while large sums of money continue to be squandered in the name of development in ways which often hurt those whom they are supposed to be helping.

Thus it would be wrong to overstate the current importance of most of the initiatives in this book and the wider movements they represent. As the first two chapters make clear, most of the major global trends are still going in the wrong direction, some at an accelerating rate. Most of the initiatives themselves are still very fragile. Even the larger ones, especially in developing countries, could all too easily be swept away by the sort of social, political and economic instability which is so common in the Third World, and may well become more so elsewhere.

As much of the earlier argument has shown, this instability is not simply an unfortunate, remediable byproduct of an otherwise beneficent power structure. Rather it is an inevitable result of an economic order which, since the earliest days of the industrial revolution, has in broad terms divided the affected population into three distinct and these days more or less equally sized groups: those who benefit from the order (industrialists and elite workers, professional middle classes); those who are dependent on it and, in effect, become its servants, with little remuneration and less control over the direction of their own lives; and those who are surplus to its requirements, the 'disposable people', the Narmada Valley oustees, the 50 per cent of some developing country populations who have not only been bypassed by mainstream development but are being systematically impoverished by it, or the part of the population of the 'free market' industrial countries, up to 20 per cent in the US and UK, who languish fixedly below the poverty line. These disposable people are not a new phenomenon of capitalist industrialisation, rather they are one of its most enduring features, discernable at least since the Enclosure Acts in Britain from the seventeenth century when the process began of depriving peasants of their independent means of subsistence and forcing them into wage labour or penury. (Disposable people have also been a feature of state socialism as well. It should be clear that no simple capitalism-to-socialism transformation is being advocated here).

There are three immensely powerful forces which are tending to reinforce this division of society into three groups – beneficiaries, servants and the dispossessed. These forces I am here calling scientism, developmentalism and statism. [I owe several of the following ideas to a most stimulating meeting in Mexico of the International Group on Grassroots Initiatives in January 1990 and especially to D.L. Sheth from the Centre for the Study of Developing Societies in Delhi, who participated in it]. By scientism I mean the hegemony of modern science and, beyond this, the attempted monopoly of the Western scientific mode of knowledge. Fritjof Capra has described how Western science has changed since medieval times:

> The medieval outlook changed radically in the sixteenth and seventeenth centuries. The notion of an organic, living and spiritual universe was replaced by that of the world as a machine, and the world-machine became the dominant metaphor of the modern era. . . . From the time of the ancients the goals of science had been wisdom, understanding the natural order and living in harmony with it . . . the basic attitude of scientists was ecological. . . . Since Bacon the goal of science has been knowledge that can be used to dominate and control nature.
>
> (Capra 1983 pp.38, 40)

Modern science is used to dominate people as well as nature. Its elevation and promotion as the *only* valid way of knowing the world dismisses the major part of the creativity, intuition and tacit and traditional knowledge, that comprise the principal perceptive, expressive and cognitive powers of most people; in short, it devalues the ideas, experience and accumulated wisdom of the majority of humankind. Indigenous systems of health care, medicine, education and agriculture, as well as ways of understanding the world and the place of people in it, have all been subject to the relentless onslaught of the modern scientific worldview. Moreover, this worldview has even failed to keep up with the advance of science itself, embodied in the post-Einsteinian scientific thinking, which emphasises uncertainty, subjectivity, cognition and the subordination of individuality to relationship. Instead the thrust of scientism has been technocratic, mechanistic, deterministic, materialistic, atomistic and anti-ecological.

While appreciating the achievements of Western science, Willis Harman has forcefully expressed the need now to look beyond it:

> The scientific view has been, in its way, outstandingly successful – yielding both technological and predictive

successes – and hence has gained tremendous prestige. It has been broadly accepted as the nearest we can come to a 'true' picture of knowledge. But it is nonetheless also true that the cosmos described by modern science is devoid of meaning and largely lacks relationship to the profound spiritual insight of thousands of years of human experience.

. . .

Few would gainsay the accomplishments of reductionistic science. For the purpose toward which it evolved – prediction, manipulation and control of the physical environment – it is superb. The issue is whether it needs to be complemented by another kind of science that can deal more adequately with wholes, with living organisms, and particularly with human consciousness.

. . .

It is impossible to create a well-working society on a knowledge base that is fundamentally inadequate, seriously incomplete, and mistaken in basic assumptions. Yet that is precisely what the modern world has been trying to do.

. . .

It begins to look as though all of the global dilemmas of which we have so recently become aware . . . [what this book has called the global problematique – PE] . . . in the end stem from our modern Western picture of reality, which we have equated with the 'reality' of reductionism.

(Harman 1988)

Elsewhere in Harman's remarkable article he specifies more particularly how he feels science could and should be restructured, emphasising participatory research; the recognition of a hierarchy of sciences, moving upwards from physical to life to human to spiritual sciences, with different methodologies appropriate for each and mechanisms of downward as well as upward causation; and a redefinition of the scientifically possible to include the total experience of humanity.

The second force, developmentalism, can be seen to derive from scientism and is the motive power of the monolithic model of development, expressed in such words as industrialisation, modernisation, consumerism, growth, etc., and measured by monetary aggregates. Developmentalism defines the principal social objectives of all countries as consumption and accumulation. These objectives are promoted by two complementary strategies. The first, from which it derives its power, can be described as the carrot of consumerism, through which it manages to create a system of total demand. Ancient cultural norms and value systems crumble before the images and artefacts of the consumer society. Even those who have no objective chance whatever of benefiting from it, who have indeed got everything to lose (i.e. the disposable people) nevertheless tend to capitulate to its perception of reality.

The minority who would resist the consumerist carrot are treated to the competitive stick of enforced economic participation. The resources and social structures that give independence or relief from the market are ruthlessly assaulted or sequestered; families and communities are ruptured; and water and biomass expropriated, not in the name of oppression but with the justification of economic efficiency and wealth creation. In a bizarre and profoundly irrational piece of sophistry, it is often claimed that those who are impoverished and immiserised by the forces of 'development' are actually (or will be imminently) its beneficiaries through some 'trickle down' process whereby some portion of the resources taken from them will be returned in more modern form. At the start of the Fourth UN Development Decade, it is clear that such claims are a cruel deception for the majority of people to whom they relate.

However, it is unlikely that the consumerist carrot or competitive stick could by themselves have wrought the transformation in the name of development which we have witnessed over the past four decades. For that the third force of statism was required, the 'sovereignty' of the nation-state, which amounts to the legitimised exercise of omnipotence over the lives of its subjects.

Competition can only rule where there are markets, which in turn depend on the allocation by price of commodities produced for exchange. Where resources are controlled by community structures and goods are produced principally for self-consumption, as was the case in most rural areas of non-industrial countries and therefore was the way of life of most of humanity until relatively recently, competition is impotent. Markets might have corroded such community self-reliance in due course, but nothing like as fast as the wholesale redefinition of property rights, effected by the nation-state in favour of exchange-oriented elites, that has occurred, especially in those countries most under US influence such as Brazil and the Philippines, in a global replay of the English Enclosure Acts of the seventeenth century which deprived the common people of their means of livelihood.

It is often forgotten that the notion of a nation-state is an extremely recent one for most of the 'sovereign' governments which now comprise the United Nations. The great majority of them date from the post Second World War period and many have been constructed along boundaries which may have suited the bureaucracies or real-politik of colonialism but which have little to do with the human realities to be governed.

The results of endowing these awkward, artificial political entities with absolute power over the people within their boundaries has been disastrous for hundreds of millions of their citizens. As the statistics quoted earlier show, many national governments have been supremely irresponsible in the conduct of their affairs: wasting their substance with massive arms spending, prestige projects and luxurious lifestyles; using those arms both to attack their neighbours and repress and torture their own citizens; ruthlessly enforcing the dominant development model on their poor people in order to finance their projects and lifestyles; and laying waste their natural environments. The most dramatic recent example of the excesses of 'sovereignty' was Saddam Hussein's renewed assault on Iraq's Kurds following the Gulf War.

None of these activities is the sole prerogative of the rulers of nation-states, of course. Despots have behaved similarly from time immemorial. But the concept of the nation-state, whose rulers have legitimate power over their subjects, allied to modern security, information and industrial technologies, has given the nation-state a penetrating power unknown to earlier tyrants. The world over, in Sarawak, Brazil, India and many other countries – peoples who have lived largely free from outside interference for millennia are now feeling the oppressive, often genocidal, impact of state power.

The promise of statehood, as of development, has thus often proved a cruel deception. So much more was expected of it than tyranny. The Charter of the United Nations, for example, began with the ringing phrase of 'We the peoples of the United Nations' and then interpreted this as 'We the governments . . .', clearly assuming that the governments and their people had something in common. The forty intervening years have proved the assumption to be deeply questionable. Time and again the independence movements which have succeeded in mobilising the people to throw off some foreign domination have coalesced into or themselves been replaced by a new tyrannic force spuriously legitimised by state sovereignty.

These, then, are the three central forces behind the modern project: *scientism* – the belief, essentially formulated in the nineteenth century, that the scientific worldview is the only valid way of perceiving and understanding the world; *developmentalism* – the belief that economic and, indeed, human progress as a whole depends on an expanding consumer society; and *statism* – the belief that the nation-state is the ultimately legitimate form of political authority. As well as creating an explosion of commodities, technologies and material expectations, these beliefs have brought humanity to the brink of wars, repression, poverty and environmental collapse of a potentially terminal nature. And it is in their practical foundation of antitheses to these beliefs that the hope for the future engendered by the initiatives described in this book resides.

Thus, in place of scientism, these initiatives celebrate the knowledge and wisdom of common people. Compared to some such knowledge, for example the understanding of their habitat possessed by the indigenous dwellers of tropical rainforests, Western scientific knowledge seems deeply deficient. But the traditional knowledge of other people regarded as ignorant from the modern viewpoint, peasants, women, slum-dwellers, is increasingly coming to be viewed as a cultural resource as valuable and valid in its own way as anything emanating from a laboratory.

This intrinsic value of traditional knowledge is augmented by the growing realisation that such knowledge is in fact the only possible basis for anything that might go by the name of 'development'. This development is a far cry from the industrial monolith with its monetary aggregates described earlier. It is development in the sense of achievement of human potential, of enhancing capability, of increasing control over the human circumstances of daily life while maintaining a healthy symbiosis with the natural processes which sustain it. Such a development can only start from what people know already. From this foundation their knowledge can, of course, be extended and enriched, by the scientific and other insights of experts, among others, but it will only remain effective knowledge for as long as it is rooted in the culture and experience of those who are developing.

For knowledge, however intrinsically relevant and effective, to be productive, there must exist the freedom and access to the necessary resources for it to be deployed. Herein lies the challenge of these initiatives to the nation-state. They can only yield their fruit through processes of participation and democracy, which demand not the abolition of the state but its transformation. State power has a vital role to play in people's self-development. It must provide the basic institutions to encapsulate and frame the market so that the market mechanism may work to general advantage. It must guarantee continuing access for all people to the resources for production and development, both monetary and non-monetary in nature. And it must implement basic norms of social justice which narrow differentials in society by progressively enabling the disadvantaged to provide for their own needs from their own resources and participate fully in its mainstream life.

It is the people's initiatives themselves which are seeking to push the state painfully towards this transformation, which is a prerequisite for a fruitful interaction between them. Repression by the state of these initiatives either kills them or hardens them into perhaps violent confrontation. Co-option by the state quenches their dynamism. The achievement of their potential by these people's processes, their very *power*, depends on their continuing autonomy and independence in a supportive social context.

This is a perplexing situation for those trapped in such notions as the 'capture' of power. Capturing the power of people's organisations ensures that either it will evaporate or will be put to other uses than those for which it was created. In the latter case it transforms the power of people over themselves into power over other people. Avoiding such a situation depends on the continuing maintenance and creation of political and social spaces within which people's power can be effectively exercised.

Thus at a fundamental level the common thrust of the new movements for social and economic change represented by those described in these pages can be expressed by the word *democratisation*: democratisation of knowledge, democratisation of development, democratisation of the state. It is a democratisation based on a new articulation of people-to-people and people-to-nature relations, which provides for, indeed promotes:

- Cultural diversity within a global perspective, rather than Western industrial hegemony;
- An ecocentric perception which places humanity within and as part of nature, rather than as external and superior to it;
- The development of people in the round, both as individuals and members of social communities, a development which caters for their needs of being, doing and relating, as well as for the now dominant consumerist need of having;
- And a mode of governance which promotes autonomy, initiative and capability, based on a commitment to social justice.

From this perspective it is clear how the battle lines for the future are drawn. On one side are scientism, developmentalism and statism, backed by the big battalions of the establishment: modern technology, and the institutions of world capitalism and state power. On the other are the people, principally the 30 per cent of humanity that is disposable as far as the modern project is concerned, but aided and abetted by many from the other 70 per cent who regard this project as ethically, socially and environmentally intolerable. The organisation of this people's power into a force that can both resist the big battalions and further the common good is a formidable and uncertain task, but this book has shown that not inconsiderable successes are possible.

Expressed like this, the big battalions against the people, there would seem to be little new in the situation and in many ways there is not. The story now, as it has always been, is partly one of resistance against exploitation and the struggle for justice. But today there is a crucial difference, because for the first time the big battalions have the power not just to destroy the people, which they have always had; nor is their reach limited to the present. Now it is also the future, the long-term future, which is at stake, and life on planet earth itself. The big battalions are fast embarking on a future which simply does not work. It remains to be seen whether the people's alternative, as outlined in this book, a future that works through a new world order, will prevail.

APPENDIX 1

THE RIGHT LIVELIHOOD AWARD AND ITS RECIPIENTS, 1980–90

War and the arms race, poverty and unemployment, resource depletion and environmental degradation, human repression and social injustice, inappropriate technologies and potent scientific knowledge untempered by ethics, cultural and spiritual decline: the Right Livelihood Award was introduced in 1980 to honour and support work which squarely faced these problems and which pioneered solutions to them. Since then forty-four people and organisations have received Right Livelihood Awards, chosen from over 300 nominations from some fifty different countries. The collective message of these initiatives is one of hope and reassurance. Today's problems are not insoluble, nor are their solution beyond the resources of individuals and small groups of people acting locally and collectively, mobilising the energies and talents of others and working for the common good. An important purpose of the Award is to project this message, in addition to supporting the initiatives themselves and disseminating the important knowledge and experience they embody.

The Right Livelihood Awards are presented annually in the Swedish Parliament in Stockholm on the day before the Nobel Prize presentations. Alfred Nobel wanted to honour those who 'during the past year have conferred the greatest benefit on mankind'. In the same spirit, the Right Livelihood Awards aim to support those working on practical and exemplary solutions to the real problems facing us today.

The Awards are presented by the Right Livelihood Awards Foundation, a charity registered in Sweden and with representation in the UK, India and the US. It is not associated with any political or religious group.

The idea and original funding came from Jakob von Uexkull, a

Swedish-German writer and philatelic expert, who is Chairman of the Foundation. Today the Award is funded in part from endowment income and by donations from individuals all over the world. The yearly cash award is, in 1990, US$120,000, which is shared by several recipients. The cash is for a specific project, not for personal use. An Honorary (non-monetary) Award is also presented to a person whose work the jury wishes to honour and recognise but who is not primarily in need of monetary support.

THE AWARD RECIPIENTS
(1980–90)

1980 **Hassan Fathy** (Egypt) for saving and adapting traditional knowledge and practices in building and construction for and with the poor.

   Plenty International (USA, Guatemala, Lesotho), a relief organisation for caring, sharing and acting with and on behalf of those in need at home and abroad.

1981 **Mike Cooley** (UK) for designing and promoting the theory and practice of human-centred, socially-useful production.

   Bill Mollison (Australia) for developing and promoting the theory and practice of permaculture – integrated low energy impact, high yield, organic growing systems.

   Patrick van Rensburg/Foundation for Education with Production (Botswana, Zimbabwe) for developing replicable educational models for the Third World majority.

1982 **Eric Dammann/Future in Our Hands** (Norway) (Honorary Award) for challenging the values and lifestyles of the West to promote a more responsible attitude towards the environment and the Third World.

   Anwar Fazal/Consumer Interpol (Malaysia) for fighting for the rights of consumers and helping them to do the same.

   Petra Kelly (West Germany), co-founder of the German Greens, for forging and implementing a new vision uniting ecological concerns with radical disarmament, social justice, feminism and human rights.

211

Participatory Institute for Development Alternatives (Sri Lanka) for developing processes of self-reliant, participatory development in the Third World.

Sir George Trevelyan/Wrekin Trust (UK) for educating the adult spirit to a new non-materialistic vision of human nature.

1983   High Chief Ibedul Gibbons (Palau, Pacific) for upholding the democratic constitutional right of Palau to be nuclear-free.

Leopold Kohr (Austria) (Honorary Award) for his early inspiration of the movement for a human scale.

Amory and Hunter Lovins/Rocky Mountain Institute (USA) for pioneering soft energy paths for global security.

Manfred Max-Neef/CEPAUR (Chile) for revitalising small and medium sized communities, fostering self-confidence and reinforcing the roots of the people.

1984   SEWA – Self-Employed Women's Association/Ela Bhatt (India) for helping home-based producers organise for their welfare and self-respect.

Winifreda Geonzon/FREE LAVA – Free Legal Assistance Volunteers' Association (Philippines) for giving assistance to prisoners and aiding their rehabilitation.

Iman Khalifeh (Lebanon) (Honorary Award) for inspiring and organising the Beirut peace movement.

Wangari Maathai/Green Belt Movement (Kenya) for converting the Kenyan ecological debate into mass action for reforestation.

1985   Cary Fowler (USA) and Pat Mooney (Canada)/Rural Advancement Fund International for helping the Third World preserve its genetic plant resources.

Lokayan (India) for linking and strengthening local groups working to protect civil liberties, women's rights and the environment.

Theo van Boven (Netherlands) (Honorary Award) for

speaking out on human rights abuse in the international community without fear or favour.

Duna Kör/Janos Vargha (Hungary) for working under unusually difficult circumstances to preserve a vital part of Hungary's environment.

1986 **Rosalie Bertell** (Canada) for raising public awareness about the destruction of the biosphere and human gene pool, especially by low-level radiation.

Alice Stewart (UK) for bringing to light in the face of official opposition the real dangers of low-level radiation.

Robert Jungk (Austria) (Honorary Award), an indefatigable fighter for sane alternatives and ecological awareness.

Ladakh Ecological Development Group (India) for preserving the traditional culture and values of Ladakh against the onslaught of tourism and development.

Evaristo Nugkuag/AIDESEP (Peru) for organising to protect the rights of the Indians of the Amazon Basin.

1987 **Chipko Movement** (India) for its dedication to the conservation, restoration and ecologically-sound use of India's natural resources.

Hans-Peter Dürr/Global Challenges Network (West Germany) for his profound critique of SDI and work to convert high technology to peaceful uses.

Johan Galtung (Norway) (Honorary Award) for his systematic and multidisciplinary study of the conditions which can lead to peace.

Institute for Food and Development Policy/Frances Moore Lappé (USA) for revealing the political and economic causes of world hunger and how citizens can help to remedy them.

Mordechai Vanunu (Israel) for his courage and self-sacrifice in revealing the extent of Israel's nuclear weapons programme.

1988 **International Rehabilitation and Research Centre for**

Torture Victims/Inge Kemp Genefke (Denmark) (Honorary Award) for helping those whose lives have been shattered by torture to regain their health and personality.

José Lutzenberger (Brazil) for his protection and conservation of the natural environment in Brazil and abroad.

John F. Charlewood Turner (UK) for championing the rights of people to build, manage and sustain their own shelter and communities.

Sahabat Alam Malaysia/Mohamed Idris, Harrison Ngau, the Penan people (Malaysia) for their exemplary struggle to save the tropical rainforests of Sarawak.

1989 **Aklilu Lemma and Legesse Wolde-Yohannes** (Ethiopia) for discovering and campaigning relentlessly for an affordable preventative against bilharzia.

The Seikatsu Club Consumers' Cooperative (Japan) (Honorary Award) for creating the most successful, sustainable model of production and consumption in the industrial world.

Survival International (UK) for working with tribal peoples to secure their rights, livelihood and self-determination.

Melaku Worede (Ethiopia) for preserving Ethiopia's genetic wealth for the benefit of all humanity.

1990 **Asociacion de Trabajadores Campesinos del Carare** (Colombia), an organisation of peasants and workers, for their commitment to peace, family and community even when surrounded and threatened by the most senseless violence.

Felicia Langer (Israel) for the courage of her struggle for the basic human rights of Palestinians in the very difficult circumstances of the Israeli military courts.

Alice Tepper Marlin/Council on Economic Priorities (USA) for showing the directions in which the Western economy must develop for the wellbeing of humanity.

Bernard Ledea Ouedraogo/Six S Association (Burkina Faso) for establishing and strengthening peasant self-help movements all over West Africa.

PUBLICATIONS
The following books about the Award have appeared:

English

People and Planet, Replenishing the Earth (Green Books, Hartland, Devon, UK, 1987, 1990), the Award speeches of 1980–85 and 1986–89 edited by Tom Woodhouse, available from Right Livelihood Awards Office (address below) £6.50, £10.00 (plus p&p) respectively.

German

Der Alternative Nobelpreis by Jakob von Uexkull (Dianus Trikont, 1986) available as above, DM 28 (plus p&p).

Projekte der Hoffnung: Der Alternative Nobelpreis by Jakob von Uexkull and Bernd Dost (Raben Verlag, München, 1990).

Italian

Il Premio Nobel Alternativo by Jakob von Uexkull (Edizioni Mediterranee, Roma, 1988).

Swedish

Vi och var jord (Energica Forlag, 1989), available from Swedish office below.

For information about the Awards and how to make nominations, contact:

Administrative Director
Right Livelihood Awards Administrative Office
PO Box 15072
S–10465 Stockholm
Sweden

Tel: (08) 702 03 40
Int. code: +46

APPENDIX 2

ADDRESSES OF PRINCIPAL ORGANISATIONS MENTIONED

CHAPTER 2

Worldwatch Institute (State of the World Reports)
1776 Massachusetts Ave NW
Washington DC 20036
USA

CHAPTER 3

END/European Dialogue
11 Goodwin Street
London EC1V 7JT
UK

Great Peace Journey
International Office
Villa Flora
Brunnsparken
S–37200 Ronneby
Sweden

Mouvement International de la Réconciliation (MIR) (Goss-Mayr)
47 Rue de Clichy
75009 Paris
France

Pugwash Office (UK)
Flat A
63a Great Russell Street
London WC1B 3BJ
UK

Saferworld Foundation
82 Colston Street
Bristol BS1 5BB
UK

Yesh Gvul
P.O. Box 6953
Jerusalem 91 068
Israel

CHAPTER 4

Amnesty International
1 Easton Street
London WC1X 8DJ

CEFEMINA
Apantado 5355–1000
San José
Costa Rica

Human Rights Internet
c/o Human Rights Center
University of Ottawa
57 Louis Pasteur
Ottawa, Ontario
K1N 6N5 Canada

HURIDOCS
Torggata 27
N–0183 Oslo 1
Norway

International Rehabilitation and Research Centre for
    Torture-Victims
Juliane Maries Vej 34
DK–2100 Copenhagen O
Denmark

Seventh Generation Fund
PO Box 10
Forestville CA 95436
USA

Survival International
310 Edgware Road
London W2 1DY
UK

Women Living Under Muslim Laws
B.P. 23
F–34790 Grabels
France

CHAPTER 5

Dag Hammarskjöld Foundation
Övre Slottsgatan 2
S–75220 Uppsala
Sweden

Movement Against Big Dams (Baba Amte)
Anandwan
Dist Chandrapur
Maharashtra 442914
India

Sarvodaya Shramadana Movement
98 Rawatawatte Road
Moratuwa
Sri Lanka

World Bank
1818 H Street
Washington DC 20433
USA

CHAPTER 6

Bangladesh Rural Advancement Committee
66 Mohakhali CA
Dhaka–1212
Bangladesh

Council on Economic Priorities
30 Irving Place
New York NY 10003
USA

Grameen Bank
Mirpur Two
Dhaka–1216
Bangladesh

International Federation for Alternative Trade
PO Box 2703
1000 CS Amsterdam
Netherlands

International Organization of Consumers' Unions (IOCU)
Regional Office for Europe and North America
Emmastraat 9
2595 EG The Hague
Netherlands

New Consumer
52 Elswick Road
Newcastle upon Tyne NE4 6JH
UK

Seikatsu Club Consumers' Cooperative
2–26–17 Miyasaka
Setagaya-ku
Tokyo 156
Japan

Six S Association/NAAM Movement
BP 100
Ouahigouya
Burkina Faso

South Shore Bank
71st and Jeffery Boulevard
Chicago
Illinois 60649–2096
USA

Third World Information Network
(Pauline Tiffen)
345 Goswell Road
London EC1V 7JT
UK

Working Women's Forum
55 Bhimasena Garden Road
Mylapore
Madras 600004
India

CHAPTER 7

Asociación ANAI
Apartado 902
Limon 7300
Costa Rica

AUGE
Christian-Foerster-Str. 19
2000 Hamburg 20
Germany

BAUM
Tinsdaler Kirchenweg 211
2000 Hamburg 56
Germany

Chipko Movement
Parvatiya Navjeevan Mandal
Silyara
Tehri-Garhwal UP
Via Ghansali
Pin 249155
India

Environmental Defense Fund
1616 P Street NW
Washington DC 20036
USA

Green Belt Movement
Ragati Road
PO Box 43741
Nairobi
Kenya

Rocky Mountain Institute
Old Snowmass
Colorado 81654
USA

Sahabat Alam Malaysia
43 Salween Road
10050 Penang
Malaysia

CHAPTER 8

Athenian School of Homeopathic Medicine
1 Perikleous str.
15122 Maroussi
Athens
Greece

Foundation for Education with Production
POB 20906
Gaborone
Botswana

Gesundheit Institute
2630 Robert Walker Place
Arlington VA 22207
USA

Gonoshasthaya Kendra
Dhaka–1350
PO Nayarhat
Bangladesh

3Hesperian Foundation
PO Box 1692
Palo Alto
California 94302
USA

Lightmoor Project
Chapel House
7 Gravel Leasowe
Lightmoor
Telford
Shropshire

New School Programme
Calle 30a No 22–57
Bogota
Colombia

Orangi Pilot Project
1D/26 Daulat House
Orangi Town
Karachi–41
Pakistan

Swadhyaya
c/o Vimal Jyoti
6/8 Dr Wilston St
V.P. Road
Bombay 400004
India

Village Action Movement
Suomen Kunnallisliitto Kehittamisosasto
Albertinkatu 34c
00180 Helsinki
Finland

CHAPTER 9

Institute of Noetic Sciences (Harman)
Suite 300
475 Gate Five Road
Sausalito
California 94965
USA

BIBLIOGRAPHY

Adams, H. 1988 'A Washington life: a doctor who refuses to charge his patients', *Washington Post*, 14 February 1988, pp.20–3.

Adams, R., Carruthers, J. and Hamil, S. 1991 *Changing Corporate Values*, Kogan Page, London.

Agarwal, A. 1985 'The state of the environment and the resulting state of "the last person"', 5th Annual World Conservation Lecture, World Wide Fund for Nature UK, Godalming, Surrey.

Allen, A. and G. 1988 *Everyone Can Win: Opportunites and Programs in the Arts for the Disabled*, EPM Publications, McLean VA.

Alternative Defence Commission 1983 *Defence Without the Bomb*, Taylor & Francis, New York.

Alvares, C. 1989 'No!', *The Illustrated Weekly of India*, 15–21 October 1989.

Amnesty International 1989 *Amnesty International Report 1989*, Amnesty International Publications, London.

Amnesty International, undated 'Your letter can save lives', Amnesty International leaflet, London.

Amte, B. 1989 *Cry, the Beloved Narmada*, Maharogi Sewa Samiti, Anandwan, Maharashtra, India.

Amte, B. 1990 *The Case Against Narmada and the Alternative Perspective*, Amte, Anandwan, India (address in Appendix 2).

Apitz, K. and Gege, M. 1990 *Was Manager von der Blattlaus Lernen Können*, Gabler Verlag, Wiesbaden (English translation forthcoming).

Appleton, B. 1984 'Dams dammed', *The New Civil Engineer*, 15 November 1984, London.

Ariyaratne, A. 1985 *Sarvodaya Shramadana Movement (SSM) of Sri Lanka: a Case Study*, SSM, Moratuwa, Sri Lanka.

ASHM (Athenian School of Homeopathic Medicine) undated *George Vithoulkas: the Man, His Works, His Vision*, ASHM, Athens.

AUGE (Aktionsgemeinschaft Umwelt, Gesundheit, Ernahrung) 1986 *Feasibility Demonstration Project for Household and Community Environmental Advisers*, Commission of European Communities Contract No.6613(86)01, AUGE, Hamburg.

Barratt Brown, M. 1988 'Introductory address' in *Who Cares About Fair*

Trade?, Report of a conference on 'Development, Trade and Cooperation', September 1988, TWIN (Third World Information Network), London.

Beckman, D. 1987 'Environmental concern about Singrauli', office memorandum, 11 March 1987, World Bank, Washington DC.

Bhushan, M. 1989 'Vimochana: women's struggles, non-violent militancy and direct action in the Indian context', *Women's Studies International Forum*, vol. 12, no. 1, pp.25–33.

Booth, K. *et al.* 1990 *European Security: the New Agenda*, Saferworld Foundation, Bristol.

BRAC (Bangladesh Rural Advancement Committee) 1988 *A Brief on BRAC*, BRAC, Dhaka, Bangladesh.

BR1–ICIDI (Independent Commission on International Development Issues) 1980 *North South: a Programme for Survival*, Pan, London.

BR2–Brandt Commission 1983 *Common Crisis*, Pan, London.

Brock-Utne, B. 1989 'Women and Third World countries – what do we have in common?', *Women's Studies International Forum*, vol. 12, no. 5, pp.495–503.

Brown, L. *et al.* 1986 *The State of the World 1986*, Norton, New York.

Brown, L. *et al.* 1989 *The State of the World 1989*, Norton, New York.

Brown, L. *et al.* 1990 *The State of the World 1990*, Norton, New York.

CAAT (Campaign Against Arms Trade) 1981 'British military involvement in Argentina', CAAT Factsheet no.32, CAAT, London.

Caldicott, H. 1985 *Missile Envy: the Arms Race and Nuclear War*, Bantam, New York.

Capra, F. 1983 *The Turning Point: Science, Society and the Rising Culture*, Fontana, London, (first published 1982 by Simon & Schuster, New York).

Carrol, B. 1989 'Women take action! Women's direct action and social change', *Women's Studies International Forum*, vol. 12, no. 1, pp.3–24.

CEA (Council of Economic Advisers) 1990 *Annual Report to the President*, CEA, Washington DC.

CEP (Council on Economic Priorities) 1988 *Shopping for a Better World*, CEP, New York.

CEP (Council on Economic Priorities) 1990 *Environmental Data Clearinghouse*, CEP, New York.

Chalmers, M. 1990 'The peace dividend: a European perspective' in Booth *et al. European Security: the New Agenda*, Saferworld Foundation, Bristol.

Chambers, R. 1985 *The Working Women's Forum: a Counter-Culture by Poor Women*, UNICEF, New York.

Chambers, R. 1988 *Sustainable Livelihoods, Environment and Development: Putting Poor Rural People First*, Institute for Development Studies (IDS) Discussion Paper 240, IDS, University of Sussex, Brighton.

Chowdhury, Z. 1989 'Delivering the services, primary health with the rural community: a case study', presented at ICOMP workshop November 16/17, mimeo, Gonosthaya Kendra, Dhaka, Bangladesh.

Coates, J. and Jarratt, J. 1989 *What Futurists Believe*, Lomond, Mount Airy MD.

Colbert, V. 1987 'Universalisation of primary education in Colombia: the New School Programme', mimeo, Bogota, Colombia.

Colchester, M. and Lohmann, L. 1990 *The Tropical Forestry Action Plan: What Progress?*, World Rainforest Movement/Friends of the Earth/*The Ecologist*, Penang and London.

Conable, B. 1987, speech to World Resources Institute, Washington DC, 5 May 1987.

Corry, S. 1989 *Address in London on Receiving The Right Livelihood Award 1989* (on behalf of Survival International), Survival International, London.

Cox, B.S. 1987 'Thailand's Nam Choam Dam: a disaster in the making', *The Ecologist*, vol.16, no.6, pp.212–20.

Dag Hammarskjöld Foundation (DHF) 1975 *What Now? Another Development*, DHF, Uppsala.

Dag Hammarskjöld Foundation (DHF) 1977 *Another Development: Approaches and Strategies*, DHF, Uppsala.

Daly, H.E. and Cobb, J.B. 1990 *For the Common Good: Redirecting the Economy Towards Community, the Environment and a Sustainable Future*, Merlin Press, London (first published 1989, Beacon Press, Boston).

DEF (Department of Environment and Forests) 1987 *Environmental Aspects of Narmada Sagar and Sardar Sarovar Multi-Purpose Projects*, DEF, Delhi.

De-Silva, G., Haque, W., Mehta, N., Rahman, A. and Wignaraja, P. 1988 *Towards a Theory of Rural Development*, Progressive Publishers, Lahore, Pakistan.

DOE (Department of the Environment) 1988 *Our Common Future: a Perspective by the United Kingdom on the Report of the World Commission for Environment and Development*, DOE, London.

DOE (Department of the Environment with the Department of Trade and Industry) 1989 'Environmental labelling: a discussion paper', August 1989, DOE, London.

Dürr, H-P. 1986 'Proposal for a world peace initiative', endorsed by Council of International Physicians for Prevention of Nuclear War, June 1986, Köln.

Eavis, P. (ed.) 1990 'Executive summary of European security: the new agenda', pamphlet, Saferworld Foundation, Bristol.

Eavis, P. and Clarke, M. 1990 *Security after the Cold War: Redirecting Global Resources*, Saferworld Foundation, Bristol.

Economist 1990 'Survey South Africa', *The Economist*, London, November, p.12.

EDF (Environmental Defense Fund) 1985 'Environmental Defense Fund spurs law for environmental reform of World Bank lending', EDF Press Release, December 1985, EDF, Washington DC.

EDF (Environmental Defense Fund) 1987 'The failure of social forests in Karnataka', *The Ecologist*, vol.17, no.4/5, pp.151–4.

EDF (Environmental Defense Fund) undated 'The Polonoroeste campaign in Brazil: a case history', Appendix B (to unpublished document), EDF, Washington DC.

Einarsson, A. 1989, speech at a Nordic Peace Zone seminar, May 1989, Great Peace Journey, Uppsala.

BIBLIOGRAPHY

Ekins, P. 1986 *The Living Economy: a New Economics in the Making*, Routledge & Kegan Paul, London.
Ekins, P. 1989 'Trade and self-reliance', *The Ecologist*, vol.19, no.5 pp.186–90, September/October.
Ekins, P. 1992 *Real Life Economics: The Emergence of a Living Economy School of Thought*, Routledge, London (forthcoming).
ELC (Environment Liaison Centre) 1986 'NGOs and environment-development issues', paper submitted to the World Commission on Environment and Development, ELC, Nairobi.
Elkington, J. and Hailes, J. 1988 *The Green Consumer Guide*, Victor Gollancz, London.
Engineer, R. 1989 'The Sardar Sarovar controversy: are the critics right?', a special report in *Business India*, 30 October–12 November 1989, pp.90–104.
EOC (Equal Opportunities Commission) 1989 'The fact about women is . . .', leaflet, EOC, Manchester.
European Dialogue (ED) 1990, Information Brochure, ED, London.
FAO (Food and Agriculture Organization) 1987 *The Tropical Forestry Action Plan*, FAO, Rome.
Gaia Foundation 1989, circular letter dated December 1989 giving details of rainforest conservation efforts from around the world, Gaia Foundation, London.
Galtung, J. 1967 *Theory and Methods of Social Research*, George Allen & Unwin, London.
Galtung, J. 1985 'Twenty-five years of peace research: ten challenges and some responses', *Journal of Peace Research*, vol.22, no.2, pp.141–58.
Galtung, J. 1987 'Visioning a peaceful world', speech on receiving the 1987 Right Livelihood Award, published in Woodhouse, T. (ed.) 1990 *Replenishing the Earth: the Right Livelihood Award Speeches 1986–89*, Green Books, Hartland, Devon.
Gege, M., Jungk, H., Jurgen-Pick, H. and Winter, G. 1986 *Öko-Sparbuch für Haushalt und Familie*, Mosaik-Verlag, München.
George, S. 1988 *A Fate Worse Than Debt*, Penguin, London.
GI (Gesundheit Institute) 1987 *Gesundheit Institute: a modern community addressing the major problems in health care delivery*, GI, Arlington VA.
Goldsmith, E. and Hildyard, N. 1984, 1986 *The Social and Environmental Effects of Large Dams*, vol.1 1984, vol.2 1986, Wadebridge Ecological Centre, Wadebridge, Cornwall.
Gorbachev, M. 1988, Address to United Nations General Assembly, 12 December, United Nations, New York.
Goulet, D. 1981 *Survival With Integrity: Sarvodaya at the Crossroads*, Marga Institute, Colombo, Sri Lanka.
GPJ (Great Peace Journey) 1987 *Five Questions to the Governments and their Answers*, Report to the United Nations, second edition, May 1987, GPJ, Uppsala.
Grameen Trust 1989 *Grameen Dialogue*, vol.1, no.1, September 1989, Grameen Trust, Dhaka, Bangladesh.

227

Harman, W. 1988 'The transpersonal challenge to the scientific paradigm: the need for a restructuring of science', *Revision*, vol.11, no.2, Fall 1988.

Hancock, G. 1989 *Lords of Poverty*, Macmillan, London.

Hasan, A. 1986 'The low cost sanitation project of the Orangi Pilot Project and the process of change in Orangi', paper given at the seminar 'The Essence of Grassroots Participation in Human Settlements Work: an Asian Perspective', February 1986, Thailand.

Hautamaki, L. 1989 *Elävä Kylä – Kylätoiminnan Opas (Living Village – a Guide to Village Action)*, Gummerus, Jyväskylä, Finland.

Havel, V. 1988 'Anti-politics', in Keane, J. (ed.) *Civil Society and the State*, Verso, London.

Hengstenberg, J. 1988 'The history of global challenges network (GCN) projects', in *Report of the 2nd Conference on Ecology, Economy and Modern Communications Technologies*, November 1988, GCN, München.

Hewlett, S.A. 1986 *A Lesser Life: The Myth of Women's Liberation in America*, Warren Books, New York.

Houver, G. 1989 *A Non-Violent Lifestyle: Conversations with Jean & Hildegard Goss-Mayr*, Marshall Morgan & Scott, London.

Huhne, C. 1989 'Some lessons of the debt crisis: never again?', in O'Brien, R. and Datta, T. (eds) *International Economics and Financial Markets: the AMEX Bank Review Prize Essays 1988*, Oxford University Press, Oxford.

Humm, T. 1985 'The economic effects of the arms trade', paper presented to The Other Economic Summit (TOES), 17–19 April 1985, London.

HURIDOCS 1989 *Report and Evaluation of Activity, 1985–1989*, HURIDOCS, Oslo.

ICDSI (Independent Commission on Disarmament and Security Issues) 1982 *Common Security: a Programme for Disarmament*, Pan, London.

Ikanan, E.K. 1986, speech on receiving the 1986 Right Livelihood Award, published in Woodhouse, T. 1990 (ed.) *Replenishing the Earth: the Right Livelihood Award Speeches 1986–89*, Green Books, Hartland, Devon.

IUCN (International Union for the Conservation of Nature and Natural Resources) 1980 *World Conservation Strategy*, IUCN, Geneva, published in popular form as *How to Save the World*, Kogan Page, London, 1981.

IUCN 1989 *Case Studies in Population and Natural Resources: Costa Rica*, IUCN, Geneva.

Jhaveri, N. 1988 'The three gorges debacle', *The Ecologist*, vol.18, no.2/3, pp.56–64.

Journe, V. 1990 *Co-Conversion of Military-Orientated Activity to Peaceful Application*, Report of Working Group 5, 40th Pugwash Conference on Science and World Affairs, September 15–20, Pugwash, London.

Kalpavriksh 1988 'The Narmada Valley Project: a critique', mimeo, Kalpavriksh, Delhi.

Kane, R. 1990 'Clearance for Narmada lapses', *Indian Express*, 1 November 1990.

Kantowsky, D. 1988 *Learning How to Live in Peace: the Sarvodaya Movement of Sri Lanka*, SSM, Moratuwa, Sri Lanka.

Keane, J. (ed.) *Civil Society and the State*, Verso, London.

Keepin, B. 1990 'Nuclear power and global warming' in Leggett J. (ed.) *Global Warming: The Greenpeace Report* Oxford University Press, Oxford, pp.295–316.

Kelly, P. 1984 *Fighting for Hope*, Chatto & Windus, London.

Khan, A.H. 1990a, *House Building by Low Income Families in Orangi*, Orangi Pilot Project, Karachi.

Khan, A.H. 1990b, *Orangi Pilot Project (OPP) Models*, OPP, Karachi.

Knevitt, C. 1987 'Today's belt and braces winners', *The Times*, 3 July 1987, London.

Lamb, R. 1981 'Saving the arms of Lord Vishnu', UNEP (United Nations Environment Programme) Feature/33, September 1981, UNEP, Nairobi.

Leggett, J. (ed.) 1990 *Global Warming: The Greenpeace Report*, Oxford University Press, Oxford.

Lohmann, L. 1990a 'Commercial tree plantations in Thailand: deforestation by any other name', *The Ecologist*, vol.20, no.1 pp.9–17.

Lohmann, L. 1990b 'Remaking the Mekong', *The Ecologist*, vol.20, no.2, pp.61–6.

Lovins, A. 1989 'Energy, people and industrialization', paper commissioned for the Hoover Institution Conference 'Human Demography and Natural Resources', February 1989, Stanford University, CA.

Lovins, A. and H. 1986 'Building real security', speech on receiving the 1983 Right Livelihood Award, published in Woodhouse T. (ed.) 1986 *People and Planet: the Right Livelihood Speeches 1980–85*, Green Books, Hartland, Devon.

Lydenburg, S., Strub, S.O. and Tepper-Marlin, A. 1986 *Rating America's Corporate Conscience*, Addison & Wesley, New York.

McBride, B. *et al.* 1988 *Sarvodaya Shramadana Movement: a Mid-Term Review*, SSM, Moratuwa, Sri Lanka.

McLarney, W. 1985 'About ANAI', ANAI (Asociación de Los Nuevos Alquimistas), Limon, Costa Rica.

McLarney, W. 1988 'Gandoca/Manzanilla Wildlife Refuge, Costa Rica', in *Biological Conservation Newsletter*, no.63, June 1988, Smithsonian Institution, Washington DC.

MacNeill, J., Cox, J. and Runnalls, D. 1989 *CIDA and Sustainable Development*, Institute for Research on Public Policy, Halifax, Nova Scotia.

Manning, C. 1987 'Families building a better future', *Birmingham Post*, 8 July 1987, Birmingham.

Max-Neef, M., Elizalde, A. and Hopenhayn, M. 1989 'Human scale development: an option for the future', in *Development Dialogue*, 1989: 1, Dag Hammarskjöld Foundation, Uppsala.

Meadows, D.H., Meadows, D., Randers, J. and Behrens, W. 1972 *Limits to Growth*, Universe Books, New York.

Melman, S. 1986 'Limits of military power, economic and other', *International Security*, vol.11, no.1 (Summer).

Momsen, J. 1991 *Women and Development in the Third World*, Routledge, London.

Myers, N. 1986a *Tackling Mass Extinction of Species: a Great Creative*

Challenge, Horace M. Albright Lectureship in Conservation, University of California, Berkeley.

Myers, N. 1986b 'The environmental dimension to security issues,' *The Environmentalist*, vol.6, no.4, pp.251–7.

Myers, N. 1989 *Deforestation Rates in Tropical Forests and their Climatic Implications*, Friends of the Earth, London.

Navaratne, S. 1988 'Politics of the soul', in *Dana*, International Journal of the Sarvodaya Shramadana Movement, vol.XIII, nos.7 & 8, July/August 1988, Moratuwa, Sri Lanka.

NBA (Narmada Bachao Andolan) 1989 *Narmada: a Campaign Newsletter*, no.2, August 1989, NBA, Delhi.

Netherlands (Government of) 1990 *Report of the Joint Evaluation Mission of Working Women's Forum*, available from WWF, Madras, India.

Oppenheim, C. 1990 *Poverty: the Facts*, Child Poverty Action Group, London.

OPP (Orangi Pilot Project) 1988 *Documentation on the Orangi Pilot Project: volume 1 – Introduction to Programme and Methodology*, OPP, Karachi.

OPP 1989 *Quarterly Progress Report, Jan–March 1989*, OPP, Karachi.

OPP 1990 *Quarterly Progress Report, Jan–March 1990*, OPP, Karachi.

Osborne, D. 1988 *Laboratories of Democracy*, Harvard Business School Press, Cambridge MA.

Patkar, M. 1990 'A note on the actions and recommendations for Sardar Sarovar Project (SSP) stipulated by the World Bank with a deadline of May 15 1990', presented as an attachment to Rich 1990b.

Pearce, D., Markandya, A. and Barbier, E. 1989 *Blueprint for a Green Economy*, Earthscan, London.

Pearson, L.B. *et al.* 1969 *Partners in Development*, Praeger, Westport CT.

Pick, H. 1990 'PM backs down on missiles', *The Guardian*, 20 March 1990, London and Manchester.

Pietilä, H. 1989 'Village Action Movement as the only broad social movement in Finland for humanity and self-determination of people', 5 July 1989, mimeo, Helsinki.

Pietilä, H. and Vickers, J. 1990 *Making Women Matter: the Role of the United Nations*, Zed Books, London.

Pradervand, P. 1989 *Listening to Africa*, Praeger, Westport CT.

Preston, L. and Sapienza, H. 1989 'Stakeholder management and corporate performance', paper presented to Conference on Socio-Economics, April 1989, Harvard Business School, Cambridge MA.

Pugwash Council (PC) 1990 'The Pugwash Conferences on Science and World Affairs', pamphlet, PC, London.

Raj, P.A. 1989 *Facts: Sardar Sarovar Project*, speech given on 10 June 1989 in Bombay, organised by Indian Science Communicators Association.

RCT (Rehabilitation Centre for Torture Victims) 1990 *Report from the International Symposium on Torture and the Medical Profession*, RCT, Copenhagen.

Rich, B. 1987 'International Conservation Project: progress report, April/May 1987', memorandum, EDF, Washington DC.

Rich, B. 1989, 'Statement on behalf of the Environmental Defense Fund (EDF) and others concerning the environmental performance of the World Bank (before two congressional sub-committees)', 26 September, EDF, Washington DC.

Rich, B. 1990a 'The Emperor's new clothes: the World Bank and environmental reform', *World Policy Journal*, Spring, pp.305–29.

Rich, B. 1990b 'Statement on behalf of Environmental Defense Fund (EDF) and National Wildlife Federation covering the environmental performance of the public international financial institutions and other related issues before the Sub-Committee on Foreign Operations, Committee on Appropriations, US Senate', 25 July EDF, Washington DC.

RMI (Rocky Mountain Institute) 1990 'The crisis in the Persian Gulf: here we go again', *RMI Newsletter* vol.VI, no.3, Fall/Winter, RMI, Old Snowmass CO.

Robertson, J. 1989 *Future Wealth: a New Economics for the 21st Century*, Cassell, London.

Rocha, J. 'Brazil names radical to draft ecology plan', *The Guardian*, 5 March 1990, London and Manchester.

Roy, S.K. 1987 'The Bodhghdat Project and the World Bank', *The Ecologist* vol.17, no.2/3, pp.62–73.

Sachs, W. 1989 'On the archaeology of the development idea: six essays', mimeo, Science, Technology and Society Program, Penn State University, University Park PA.

SAM (Sahabat Alam Malaysia) 1984 'State of the Malaysian environment', SAM, Penang, Malaysia.

SF (Saferworld Foundation) 1990 *The Gulf Crisis: Test Case for the New World Order*, September, Saferworld Foundation, Bristol.

Schneider, B. 1988 *The Barefoot Revolution: a Report to the Club of Rome*, Intermediate Technology Publications, London.

Seikatsu Club 1988 *The Seikatsu Club: Autonomy in Life*, Seikatsu Club, Tokyo.

Seventh Generation Fund (SGF) 1988 *Rebuilding Native American Communities: Seventh Generation Fund Annual Report 1987/88*, SGF, Lee NV.

Shapiro, J. 1989 'Shorebank Corporation', paper presented to Conference on Unemployment and Consumer Debts, September 1989, Hamburg.

Siiskonen, P. 1986 'Participation of farm-families in the decision-making process in rural communities', paper given to the Third Session of the Working Party on Women and the Agricultural Family in Rural Development, October 1986, FAO/ECA, Rome.

Sivard, R. 1986 *World Military and Social Expenditures 1986*, 11th Edition, World Priorities, Washington DC.

Sivard, R. 1987 *World Military and Social Expenditures 1987–88*, 12th Edition, World Priorities, Washington DC.

Sivard, R. 1989 *World Military and Social Expenditures 1989*, 13th Edition, World Priorities, Washington DC.

Smillie, I. *et al.* 1985 *Sarvodaya Shramadana Movement: an Assessment*, SSM, Moratuwa, Sri Lanka.

Solis, V. and Trejos, M. 1990 *Women and Sustainable Development in Central America*, IUCN/CEFEMINA, San José.

Srivastava, R.K. 1989 'A short description of Swadhyaya Movement in India', mimeo, Centre for the Study of Developing Societies, Delhi.

SSM (Sarvodaya Shramadana Movement) undated 'Sarvodaya: development from village up', publicity leaflet, SSM, Moratuwa, Sri Lanka.

Steele, I. 1987 'A new way to prevent bilharzia', in *AfricAsia*, no.42, June 1987.

Survival International (SI) 1986 *Annual Report 1986/87*, SI, London.

Survival International (SI) 1989 'Resistance to dams holds promise', in *Survival International News*, no.25, SI, London.

Swedish Ministry of Foreign Affairs (SMFA) 1988 *Recovery in Africa*, SMFA, Stockholm.

Thompson, E.P. 1982 *Beyond the Cold War*, Merlin Press/END, London.

Thompson, E.P. 1987 'The year the world walked through the mirror', speech at the Final Plenary of the END Convention, Coventry Cathedral, 18 July, END, London.

Thompson, E.P. and Smith, D. (eds) 1980 *Protest and Survive*, Penguin, London.

Thompson, J. 1988 'Tribalism and the arms race trap', *Medicine & War*, vol.4, no.1, Jan–March 1988, pp.37–48.

Tiffen, P. 1990 'The latest trends in alternative trading', *Fenix*, no.00 (pilot issue), Friends of Ideas and Action Foundation, Hoofddorp, Netherlands.

Timberlake, L. 1985 *Africa in Crisis*, Earthscan, London.

Todd, J. 1987 'A short history of the ANAI Project in Talamanca', unpublished mimeo, Falmouth MA.

Tokyo Seikatsusha Network (TSN) 1988 *Building a New Japan: Seikatsusha Movement*, TSN, Tokyo.

Townsend, P. and Davidson, N. 1982 *Inequalities in Health*, the Black Report, Penguin, Markham, Ontario.

Turner, B. (ed.) 1988 *Building Community: a Third World Casebook*, Building Community Books, London.

Turner, J.F.C. 1976 *Housing by People: Towards Autonomy in Building Environments*, Marion Boyars, London.

Turner, J.F.C. 1988 'A summary of the argument for enabling policies', presented at a seminar at the University of Bradford, UK, December 1988, mimeo, Right Livelihood Award, Stockholm.

Turner, J.F.C. and Fichter, R. (eds) 1972 *Freedom to Build*, Collier Macmillan, New York.

Udall, L. 1989a 'Statement on behalf of Environmental Defense Fund (EDF) concerning the social impacts of forced resettlement in World Bank financed development projects before the Congressional Human Rights Caucus', 27 September, EDF, Washington DC.

Udall, L. 1989b 'Statement on behalf of EDF *et al.* concerning the environmental and social impacts of the World Bank financed Sardar Sarovar dam in India before the Sub-Committee on Natural Resources, Agricultural Research and Environment, Committee on Science, Space and Technology', 24 October, EDF, Washington DC.

Udall, L. 1990 'Statement on behalf of EDF *et al.* covering the environmental and social performance of the International Development Association before the Sub-Committee on International Economic Policy, Trade, Oceans and Environment, Committee on Foreign Relations, U.S. Senate', 18 July, EDF, Washington DC.

UN 1979 (Voluntary Fund for UN Decade for Women) *State of World Women 1979*, UN, New York.

UNICEF 1989 *The State of the World's Children 1989*, Oxford University Press, New York.

UNICEF 1990 *The State of the World's Children 1990*, Oxford University Press, New York.

Van Boven, T. 1989 'The international human rights agenda: a challenge to the United Nations', mimeo, University of Limburg, Maastricht.

Van Rensburg, P. 1981, untitled speech given in Stockholm on receipt of the 1981 Right Livelihood Award (RLA), December 1981, RLA, Stockholm.

Van Rensburg, P. 1984 'Activities of the Foundation for Education with Production', in *Mmegi wa Dikgang*, vol.1, no.7, 12 October 1984, Serowe, Botswana.

Waring, M. 1989 *If Women Counted: A New Feminist Economics*, Macmillan, London.

WCED (World Commission on Environment and Development) 1987 *Our Common Future*, Oxford University Press, Oxford.

Werner, D. 1977a *Where There Is No Doctor*, Hesperian Foundation, Palo Alto CA.

Werner, D. 1977b 'The village health worker: lackey or liberator?', address to International Hospital Federation Congress, May 1977, Tokyo.

Werner, D. 1985 'Public health, poverty and empowerment: a challenge', address to Johns Hopkins School of Public Health, Baltimore MD.

Werner, D. 1990 'Healthy profits in a dying world: the man-made causes of poor health', talk to Harvard School of Public Health, 25 October.

Wignaraja, P. 1988 'Participatory people-centred development', paper for the South Commission, Geneva, October.

Winter, G. 1988 *Das Umweltbewusste Management* (Business and the Environment), C.H. Beck-Verlag, München.

WLUML (Women Living Under Muslim Laws) 1987 'Zina: the Hudood Oridnance and its implications for women', Alert for Action, 21 November 1987, WLUML, Grabels, France.

WLUML 1988 *Dossier 3*, June/July, WLUML, Grabels, France.

WLUML 1989 'No deportation to death by stoning', Alert for Action, WLUML, Grabels, France.

WLUML 1990 'Open season for women', Alert for Action, 24 March 1990, based on editorial in *The Muslim*, Lahore, Pakistan, 16 March 1990, WLUML, Grabels, France.

Wong, D.A.J. 1988 'Let's first get to know the Penans', in *The People's Mirror*, 4 April 1988, Sarawak, Malaysia.

Woodhouse, T. (ed.) 1986 *People and Planet: the Right Livelihood Speeches, 1980–85*, Green Books, Hartland, Devon.

Woodhouse, T. (ed.) 1990 *Replenishing the Earth: the Right Livelihood Award Speeches, 1986–89*, Green Books, Hartland, Devon.

World Bank 1989a 'Narmada Projects and the World Bank', mimeo, World

Bank, Washington DC.
World Bank 1989b 'India – Sardar Sarovar (Narmada) Projects: background notes', mimeo, World Bank, Washington DC.
WRI (World Resources Institute) 1987 *The Tropical Forestry Action Plan: Background Information and Update*, WRI, Washington DC.
WWF (Working Women's Forum) 1988 *A Decade of the Forum*, WWF, Madras, India.
Yesh Gvul undated mimeo, Yesh Gvul, Jerusalem.

INDEX